THE
FACTORY
INSPECTORS

A LEGACY OF THE INDUSTRIAL REVOLUTION

THE FACTORY INSPECTORS

A LEGACY OF THE INDUSTRIAL REVOLUTION

EDDIE CROOKS

OBE CEng FIMechE
Formerly HM Superintending Inspector of Health & Safety

TEMPUS

Frontispiece: Photograph of Leonard Horner.

'The practice of keeping children in the mills to clean the machinery during a part of the meal times is by no means uncommon. Cotton flue or other stuff caught in the rollers, would be removed by the children, inattentive through fatigue or frivolity or both, where fingers might unwittingly be snapped by the unrelenting machine, causing ghastly mutilation or death.'
Leonard Horner. HM Inspector of Factories, 1841

First published 2005

Tempus Publishing Limited
The Mill, Brimscombe Port,
Stroud, Gloucestershire, GL5 2QG
www.tempus-publishing.com

British Library Cataloguing in Publication Data.
A catalogue record for this book is available from the British Library.

ISBN 0 7524 3569 8

Typesetting and origination by Tempus Publishing Limited
Printed in Great Britain

CONTENTS

Key to using endnote references

Works cited in this publication are listed on page 216. Each entry in the list is numbered and references within the text correspond to this list.

Acknowledgements for pictures

Many thanks to HMSO for the Core Licence (CO2W00050090) for permitting the use of illustrations from the various Factory Inspectorate publications. Thanks also to Mike Glasson for the front cover illustration.

FOREWORD

The Industrial Revolution was a time of enormous inventive genius and a golden age for eminent engineers and craftsmen and the pioneering businessmen who promoted the new manufacturing industries made possible by this creativity. The many successes of this period are well documented, both historically and in modern factual and fictional media interpretations. However, the development of the factory system did leave a dark legacy which should not be forgotten, particularly as we enter a post-industrial era and the burden of manufacture is taken up by others. The Industrial Revolution, with the development of manufacture in its many forms, demanded a high price from the multitude of men, women and, in the beginning, young children who were employed to meet its demands. Throughout this time, the human cost measured in death, injury and occupational ill health was meticulously recorded by successive Chief Inspectors of Factories in their annual reports to the government of the day. This book is a tribute to those who made this sacrifice and an acknowledgement of those who fought, often against powerful opposition, to obtain adequate legal provisions and practical solutions for the protection of those who earned their living in factories and workshops throughout the United Kingdom.

CHAPTER 1

COTTON, SLAVE TRADING AND CHILD LABOUR

The Industrial Revolution

Towards the end of the eighteenth century in the city of Manchester, Dr Thomas Percival, with other medical colleagues, formed the Manchester Board of Health to enquire into an outbreak of 'putrid fever' in the cotton mills in the region. It was found that young children employed in these factories worked long hours in squalid and dangerous conditions and were disposed to 'contagion of fever'. The Elizabethan Poor Law still governed the conditions of paupers and it was the revelation of the conditions under which pauper apprentices were shipped from the south to work in the new mills in the north that brought about the demand for reform. Recommendations were made to improve the working conditions and to provide education and moral and religious instruction for the children, and devise a new general system of laws for the wise, humane and equal government of all these works. In 1802, Sir Robert Peel the elder, himself a mill owner and an employer of some 15,000 workers, introduced a new Bill – The Health and Morals of Apprentices Act, 1802. This Act heralded the beginning of a more enlightened approach to health and welfare in mills and factories. Legislation to deal with the safety of workers was still many years away, despite the abundant evidence of dreadful accidents caused by the increased mechanisation of the Industrial Revolution.

By 1802, industrial development was well under way. In 1738 Mr John Kay of Bury invented the flying shuttle which provided a method of throwing the shuttle in a weaving loom, enabling the weaver to make twice as much cloth as before. This was improved in 1760 by his son Robert, who invented the Drop Box, allowing the weaver to use any one of three shuttles, each containing a different coloured weft. Up until this time the manufacture of cotton was a time-consuming process carried out as a cottage industry in the homes of spinners and weavers. This changed with the invention of spinning machines to match the capacity of the weaving loom. The first patent for a mechanical spinning machine was taken out by Mr John Wyatt of Birmingham in 1738. Richard Arkwright, who was born in Preston in 1732, perfected this original design and took out a patent in 1769 for the spinning machine known as the water frame, which was further improved to become the throstle. Arkwright and his partner erected a mill in Nottingham which was powered by horses, but this was an expensive way to drive the machinery and so they built a larger water-powered mill on the River Derwent, at Cromford in Derbyshire, from where the name water frame was derived.

In around 1767, a weaver named James Hargreaves of Blackburn invented a spinning machine which could spin fifteen threads simultaneously. He obtained a patent in 1770 for the spinning jenny, which by now was capable of spinning 100 threads of cotton at a time. Arkwright and Hargreaves had between them, established the genesis of the cotton industry as part of the factory system. Further improvement followed with the invention of the mule or mule jenny which combined the principles of Arkwright's water frame and Hargreaves' spinning jenny. This new machine was invented in 1779 by Samuel Crompton, a weaver from Bolton. It was recorded, 'As many as 350 hanks to the pound have been spun, each hank measuring 840 yards'.

This period of creative genius was concluded in 1785, when the Rev Dr Edmund Cartwright invented the first power loom which improved and speeded up the weaving process and restored the balance of production between spinning and weaving. The huge demand for textile machinery and an efficient means of driving the machines, the ready availability of raw materials such as iron ore and coal, combined with the strategic trading advantage of being a small island with a strong maritime tradition and a developing colonial influence with access to the American cotton trade provided a great stimulus to the Industrial Revolution, which greatly accelerated in the latter half of the eighteenth century.

The early cotton mills were powered by large water-wheels that restricted the location of the premises to sites with adequate water capacity – the Quarry Bank Mill, for example, was fitted with a water-wheel that was 32ft in diameter, providing power equivalent to 120hp – but with the introduction of James Watt's steam engine, it was no longer necessary to locate factories and mills in remote rural areas where sufficient water could be found and harnessed to drive the water-wheels. In 1812 there were over 4 million spindles in operation and by 1835 this had increased to 11 million.

Watt's invention was first patented in 1769 and further improved between 1781 and 1784. He formed a partnership with Matthew Boulton of Birmingham and installed their first steam engine at Bradley ironworks in 1782. The first use of a steam engine in the cotton industry was at Robinson's Mill in Papplewick in Nottinghamshire in 1785, where a Boulton & Watt beam engine provided the motive power to drive the textile machinery through complex assemblies of transmission shafts and belt drives. The first steam engine to be installed in a Manchester mill was in 1789 and Sir Richard Arkwright had an engine installed in his cotton mill in 1790. It was this type of mechanisation that was to prove the greatest cause of accidents in mills and factories during the next century. The use of steam, spread rapidly with the building of new factories and mills. Orrell's cotton mill at Stockport was a model of its time with 1,100 looms built on six floors, each floor being 280ft long by 50ft wide, built at a cost of £85,000.

This period in our industrial heritage was the great Age of the Machine when engineers of the day, through their inventive genius, were able to satisfy the demands

Cotton factories in Union Street, Manchester. Drawn by Austin, engraved by McGahey

of the new manufacturing industries. The machinery in the mills and factories were owned by a generation of self-made men; masters of labour who were ruthless in their determination to succeed. Apart from a few notable exceptions, the welfare of their workers was of secondary concern to the owners. They gave little thought to the demands that the new machines put upon their operators or the unhealthy working environment in many of the textile mills and factories. The rapid growth in the technology staggered the imagination and the great new factories were regarded with admiration and awe. A contemporary author, Andrew Ure, wrote the following commentary on the mills of Manchester:

> The fine spinning mills of Manchester, which have been so grossly disparaged by the partisans of the ten-hours' Bill, are, in fact, the triumph of art and the glory of England. In the beauty, delicacy and ingenuity of the machines, they have no parallel among the works of man . . . When 350 hanks are spun, containing only one pound of cotton, they form an almost incredible length of thread, extending 294,000 yards or 167 miles, and enhance the price of the material from 3s.8d to twenty five guineas . . . It is delightful to see from 800 to 1000 spindles of polished steel, advancing and receding in a mathematical line, each of them whirling all the time upon its axis with equal velocity and truth, and forming threads of surprising tenuity, uniformity and strength.[1]

It took time for the uncompromising laws of these new machines to be fully understood. The new factories were congested and working conditions were both unhealthy and dangerous. Young children were an important part of the working processes and soon became victims of the unsatisfactory working conditions. M.T. Sadler wrote:

During the whole course of the struggle that has been made on behalf of factory
it has been invariably asserted by those who have interested themselves in their ...
condition that the early and excessive labour to which they have been doomed, has not
only been rigorous to their morals and health, but in multitudes of cases, destructive of
life itself, at an age when of all others the human frame is the most tenacious of existence,
and when to destroy it by any other means than direct violence involves a degree of long
suffering and sorrow which it is distressing to contemplate even in imagination.[2]

Engels, in his observations on the working classes in 1844, noted the outward signs
of this problem in the streets of Manchester:

Besides the deformed persons a great number of maimed ones may be seen going about
Manchester, this one has lost an arm or part of one, that one a foot, the third half of a leg:
it is like living in the midst of an army returning from a campaign.[3]

In contrast to the above opinions, there were many eminent people prepared to
defend the factory system and also to question the motives of those advocating the
improvement of working conditions in the mills and factories. In 1835, Edward Baines
published a book which provided a detailed account of the cotton industry and praised
its undoubted benefits to the economy at that time. He made favourable comparisons
with working conditions in other industries and was particularly concerned about the
adverse statements being made to a committee of Factory Commissioners, appointed by
Parliament to inquire into the working conditions of children in the mills and factories:

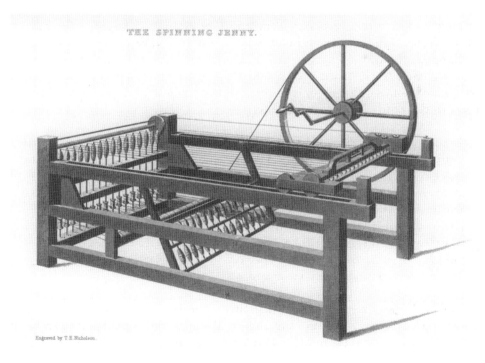

THE SPINNING JENNY.

Engraved by T. E. Nicholson.

Hargreaves' spinning jenny. Drawn by T. Allom, engraved by J. Carter

But it has been represented by declamatory writers, and even by some parliamentary orators, that the high wages of the cotton spinners are earned by the entire sacrifice of health and comfort, – that the labour of the mill is so severe, incessant, and prolonged, as to destroy the constitution and to exhaust the mental energies of the workman – that they breathe a heated and polluted atmosphere, loaded with dust and fibre of cotton, which, entering the lungs, soon produce consumption – that the exhaustion of their bodies by labour drives them to intemperance as a relief and a stimulus – that thus their lives are passed in an alternation of depressing drudgery and maddening excitement, without any healthy exercise of the mental faculties, or any rational enjoyment. It is pretended that the mill operatives are placed in cruel competition with machinery, whose relentless speed strains their faculty to the utmost, admits not of a moments intermission from toil, makes no allowance for human feebleness, but unnaturally taxes flesh and sinews to keep pace with wheels and arms of iron By those rhetoricians, the steam-engine is represented as a tyrant power, and a curse to those who work in conjunction with it. Above all, it is alleged that children who labour in mills are victims of frightful oppression and killing toil, – that they are often cruelly beaten by the spinners or over-lookers, – that their feeble limbs become distorted by continual standing and stooping, and they grow up cripples, if indeed they are not hurried into premature graves, – that in many mills they are compelled to work thirteen, fourteen, or even fifteen hours per day, – that they have no time either for play or for education, – and that avaricious taskmasters, and idle unnatural parents, feed on the marrow of these of these poor innocents. To such representations it is an appropriate finish to call the factories, as has often been done, hells upon earth.[4]

The Factory Commissioners and Factory Inspectors who visited the mills and factories at that time found reasonable grounds for concern about the working conditions of children. The Factory Inspector's survey of the cotton industry in Lancashire and Cheshire established a total workforce comprising 19,247 men, 20,962 women and 27,610 children and their average weekly wage was equal to 10s 5d. The working day was not less than twelve hours during weekdays and nine hours on Saturday. Conditions of employment for the children working in the mills became a subject of much debate and civil unrest.

A year or two ago, the subject became one of powerful agitation among the working classes of the manufacturing districts, being made so by a few individuals, who were mainly, though not altogether, influenced by humane motives, but whose imagination and feelings were much stronger than their judgements. These individuals maintained, with apparent reason, that no child ought to work more than 10 hours per day, and that the mills, which then worked eleven, twelve, and in some cases even longer, should be prevented by law from working more than ten hours. A cause in itself good, was injured by the outrageous violence and unreasonable demands of its promoters, who continually presented the most hideous caricature of the effects of factory labour . . . The latter, with few exceptions, united in the clamorous demand for a 'ten hour bill' not because they believed that children were oppressed, but because they ignorantly imagined that their own labour would be shortened by such a bill from twelve hours to ten without any reduction being made in their wages.[4]

Reports from the Factory Inspectors during this period contained many disturbing accounts of dreadful accidents to children and young persons who were exposed to many dangers in the course of being employed in these places. In the reports for the year 1842 one Inspector gives an account of an accident involving the death of a man and a young girl:

A man named Campbell, the overlooker of the room in which it happened was mending a belt which was held for him by a little girl. Another girl named Burns, 14 or 15 years old, incautiously running between them and an upright revolving shaft, got her clothes entangled with the shaft, and whilst Campbell was trying to extricate her, the girl who had been holding the belt for him, being frightened threw it down and ran away. The belt getting entangled with the teeth of the shaft, caught Campbell also and both he and little Burns were drawn up, and before the machinery could be stopped, almost crushed to pieces.[96]

The same report for 1842 mentions another fatal accident in a silk mill in Stockport:

On 29th March a man was killed in a silk mill at Stockport. He was employed in fixing a part of the machinery, his foot slipped and he fell between some cog wheels and was torn to pieces in sight of his wife.

Carding, drawing and roving cotton. The drawing shows how close children were to the dangerous moving parts of the machine. Drawn by T. Allom, engraved by J.W. Lowery

Two spinning mules, from around the 1830s, showing the machine being cleaned during operation

Early legislation

Many years passed from the beginnings of the factory system before the introduction of legislation for the protection of workers. The 1802 Act was limited to the protection of the most vulnerable, the pauper children, from the adverse consequences of their employment within the factory system. Hours of work were reduced and night work was abolished; factories were to be whitewashed and properly ventilated. A voluntary system of inspection was inaugurated and local Justices of the Peace appointed Visitors to mills and factories. The Act was, at first sight, far reaching and enlightened, giving power and authority to the appointed Visitors. But the Act was conspicuous by its lack of observance. Sir Robert Peel commented in 1815 that Visitors appointed under the Act were very remiss in the performance of their duty. Many more years passed from the enactment of the 1802 Act before effective action was taken to correct its shortcomings. The shortcomings of the 1802 legislation were acknowledged in a subsequent Act of 1833, at Section XVII of that Act:[5]

   XVII. And whereas by an Act, intituled An Act for the Preservation of the Health and Morals of Apprentices and others employed in Cotton and other Mills and other Factories, passed in the Forty-second Year of the Reign of His late Majesty George the Third, it was amongst other things provided, that the Justices of the Peace of every County in which such mill was situated should appoint yearly Two Persons not interested in or in any other way connected with such mills or Factories in such County to be Visitors . . . which Visitors so appointed were empowered and required by the aforesaid Act to enter such Factories at any Time they might think fit, and examine and report in Writing whether the same were conducted according to the Laws of the Realm, . . . And whereas it appears that the Provisions of the said Act were not duly carried into execution, and that the laws for the Regulation of the Labour of Children in factories have been evaded, partly in consequence of the Want of the Appointment of proper Visitors or Officers whose duty it was to enforce their execution.

The shortcomings of the 1802 Act were caused, in part, by the rapid development of mills and factories for the production of textiles, the rapid movement of population from rural communities to towns and cities in search of employment and the close relationship between the mill owners and the Visitors appointed to regulate them. By 1833 there were more than 3,000 workplaces under inspection by a system that was no longer acceptable to the factory reformers of the day. This growth in the production of textiles increased the demand for workers and child labour in particular continued to be exploited. This is illustrated by an advertisement that appeared in the *Manchester Chronicle* on 2 February 1811:

> PARISH APPRENTICES.
>
> *The Church Wardens of Manchester have about Two Hundred APPRENTICES, of both sexes, to put out to respectable Manufacturers, Spinners, or others. They are in general very good weavers, but their Master having become insolvent, they are now chargeable to the town. Their indentures expire at various periods.*

The story of factory legislation during these early years of the nineteenth century is one of gradual imposition of regulations and controls upon owners and manufacturers who complained that such rules would ruin their business or expose them to unfair competition from less regulated and hence cheaper foreign competition. The owners were influential and obtained strong political support for their case. But the move from rural locations to towns and cities brought the plight of factory workers to the attention of the public. An alliance of Tory and Whig MPs with progressive mill owners such as John Ward, Robert Owen and others formed the Factory Reform Movement. At the same time the movement against slavery in the Colonies was used by propagandists to equate conditions of slavery with that of the factory workers in the mills and factories. In 1830, Richard Oastler, one of the leaders in the Factory Reform Movement, likened the child workers of Bradford to Negro slaves:

> . . . The very streets which receive the droppings of an 'Anti-Slavery Society' are every morning wet with the tears of innocent victims at the accursed shrine of avarice, who are compelled (not by the cart whip of the Negro slave-driver) but by the dread of the equally appalling thong or strap of the over-looker, to hasten, half dressed, but not half-fed, to those magazines of British infantile slavery − the worsted mills in the town and neighbourhood of Bradford.[6]

The connection between the plight of child workers in the industrial towns of England and the slave trade that flourished at that time was no coincidence. They were all part of the profitable triangle of trade between the ports of Liverpool and Bristol to the coastal regions of West Africa, then to the Caribbean Islands and the southern states of America, and finally back to the ports in England. Respectable businessmen using the profits made from manufacture commissioned ships to carry the manufactured goods to the east. These goods were traded for slaves who made economic ballast for the empty ships journey to the Americas from where slaves were traded for the essential raw materials such as cotton to meet the demands of the industry.

Factory Act legislation was passed in 1819 and 1825 but it was not until 1833, after the Reform Bill of 1832, that a turning point was reached. One of the first acts of the newly reformed Parliament was to set up a Royal Commission 'to collect information in the manufacturing districts with respect to the employment of children in factories and to devise the best means for the curtailment of their labour'. This Commission very quickly made proposals for the appointment of full time Factory Inspectors and the appointment of Surgeons to provide certificates as a means of verifying the age of children in employment. Lord Ashley (later to become Lord Shaftsbury), who was one of the great social reformers of the time, presented the new Bill to Parliament. Whilst agreeing that there was a need for this Bill, the Chancellor of the Exchequer, Lord Althorp, was worried about the consequences of such reform 'since adults would be unnecessarily deprived of their opportunity of making the most of their only property, their labour'. Nevertheless the Bill was enacted on 29 August 1833 as 'An Act to regulate the Labour of Children and Young Persons in Mills and Factories of the United Kingdom.'[5] The introductory Section of the new Act shows its limitations and subsequent Sections show that the Act did not apply to the employment of children in silk mills or to children employed to carry out repairs to the machinery or premises:

> I. Whereas it is necessary that the Hours of Labour of Children and young Persons employed in Mills and Factories should be regulated, in as much as there are great Numbers of Children and young Persons now employed in Mills and Factories, and their Hours of Labour are longer than is desirable, due Regard being had to their Health and Means of Education . . . That from the First Day of January One Thousand Eight Hundred and Thirty-Four no person under Eighteen Years of Age shall be allowed to work in the Night . . . Provided always that nothing in this Act shall apply or extend to the working of any Steam or any other Engine, Water-wheel, or other power in or belonging to any Mill or Building or Machinery when used in that Part of the Process or Work commonly called fulling, roughing or boiling of Woollens, nor to any apprentices or other Persons employed therein . . .

> VII. And be it enacted . . . it shall not be lawful for any Person whatsoever to employ in any Factory or Mill as aforesaid, except in Mills for the Manufacture of Silk, any Child who shall have not completed his or her Ninth Year of Age.

> XLVII. Provided always, and be it enacted, That nothing in this Act contained shall apply to Mechanics, Artisans, or Labourers under the prescribed Ages working only in repairing the Machinery or Premises.

Section XVII of the Act, having identified the deficiencies of the 1802 Act, set out to rectify these deficiencies by enacting the appointment of the first Inspectors of Factories:

> XVII . . . be it therefore enacted That upon the passing of this Act it shall be lawful for His Majesty by Warrant under His Sign Manual to appoint during his Majesty's Pleasure Four Persons to be Inspectors of Factories . . . and such Inspectors or any of them are hereby empowered to enter any Factory or Mill, and any School attached or belonging thereto, at all Times and Seasons, by Day or by Night, when such mills or Factories are at work . . .

> XVIII. And be it further enacted, That the said Inspectors or any of them shall have Power and are hereby required to make all such Rules, Regulations, and Orders as may be necessary for the due Execution of this Act, which Rules, Regulations, and Orders shall be binding on all Persons subject to the Provisions of this Act.

If the new Act was to be more successful than the earlier legislation, it was necessary to appoint Inspectors of the highest integrity. The first four Factory Inspectors to be appointed – 'by His Majesty by Warrant under His Sign Manual' – were Leonard Horner, Robert Saunders, Robert Rickards and Thomas James Howell. These four Inspectors carried out their appointed duties within four divisions of the country. James Stuart soon replaced Rickards. Horner and Stuart became the most influential in the administration of the Act. Horner had responsibility for the Northern Division or district, which in addition to the north of England also included Scotland and Northern Ireland. In this position with his base at Manchester, he was able to take a lead in the confrontation with the textile manufacturers and their opposition to any radical reform of factory legislation.

Despite the increased powers in the Act, it had limitations and only dealt with restrictions on the employment of children under the age of eighteen years. It set a minimum age of nine for employment in certain mills and children between nine and thirteen were required to attend school for twelve hours each week. Inspectors were required to establish schools and continued in this role until the Education Act 1870 passed responsibility to the Board of Education and the local authorities. The 1833 Act did not deal with safety and Inspectors could do little to prevent accidents in mills. There was no reference to the fencing of machinery, which was a constant source of serious accidents. The Commissioners had considered the question of safety but there were strong objections from the manufacturers. During the passage of the Bill through Parliament it had been argued that the effects on the manufacturing districts would be calamitous, foreigners would be better able to compete in the English market and, in consequence, workers would be impoverished. This was not the only ground on which the manufacturers were apprehensive. It was argued that the inventive genius of the engineer was rapidly introducing innovations in the design of textile machinery and it was important that these trade secrets should be protected. There was concern that if Inspectors were granted free access they might divulge these secrets to rivals.

So the manufacturers and owners were successful in securing the omission from the 1833 Act of any requirement concerning fencing of machinery or reference to accidents. The Factory Inspectors were very concerned about their lack of powers in these matters and recorded their concerns in their reports to the Home Office. The matter came to a head following the legal case of Cotterell *v.* Stocks at Liverpool Assizes in 1840. This concerned a girl of seventeen who had been trapped by a revolving shaft. She suffered broken limbs and severe body lacerations. Her employer had deducted from her wages the sum of eighteen pence for the part of the week she was unable to work. The employer lost the case and was ordered to pay damages of £100 and £600 costs. Lord Ashley, who had been responsible for the 1833 Act, represented the girl during the legal proceedings.

It so happened that the first industrial safety legislation to reach the Statute Book did not apply to factories, but to coal mines. In 1840 a Commission was set up to consider the conditions under which children worked in mines. This was set up following the Silkstone Colliery disaster in 1838, in which twenty-six boys and girls, the youngest aged seven, were drowned when the pit flooded. The Commission also discovered that children sometimes were employed as engine men to raise and lower the hoist cages. On one occasion three miners were killed because a child of nine, whilst operating the hoist engine, turned away from his work to look at a mouse. When the Coal Mines Act of 1842 was passed, it prohibited women and children from working underground and young persons below the age of twenty-one from acting as an engine man on a hoist. The House of Lords reduced the age from twenty-one to fifteen but the Bill broke new ground by having safety within its provisions.

A further Select Committee under Lord Ashley was set up to consider the 1833 Act and in February 1841 it proposed new provisions for the safety of workers. This was fully supported by the Inspectors, who had been making strong representations for similar changes. It was proposed that the cleaning of machines in motion should be prohibited, that dangerous parts of machines should be boxed off, that upright shafts and horizontal shafts within 7ft of the floor should be totally enclosed. In response to the recommendations, the Home Secretary issued instructions to the Inspectors to obtain information on the parts of mill machinery it was practical to fence. The instruction distinguished between the steam engine and the water-wheels, the main shafting or gearing and drive belts and the manufacturing machinery that connected to the drives. Inspectors were instructed to ascertain whether parts of machinery must be cleaned in motion. In his report of 4 January 1841, Horner had this to say about cleaning machinery:

> The practice of keeping the children in the mills to clean the machinery during a part of the meal times is by no means uncommon . . . cotton flue or other stuff caught in the rollers, would be removed by the child, inattentive through fatigue or frivolity or both, where fingers might unwittingly be snapped by the unrelenting machine causing ghastly mutilation or death.[7]

The Inspectors were not all in agreement on these matters and were ever mindful of the objections of the owners and the textile machinery manufacturers. There was some difficulty in deciding whether or not it was feasible to fence machinery securely without interfering with the manufacturing process. Leonard Horner had no doubts on this issue and in his report of 1 May 1844 records the following:

The wanton and careless manner in which dangerous machinery is often left exposed which might be guarded at an expense of a few shillings and without any impediment to the working is perfectly inexcusable ... it ought to be interfered with under the authority of law.[8]

In 1844, an Act to amend the Laws relating to Labour in Factories[9] was enacted. For the first time it contained provisions for the safety of children and young persons by requiring protection from machinery and mill gearing in use in factories. The relevant Sections of the Act were as follows:

XX ... That no Child or Young Person shall be allowed to clean any Part of Mill-gearing in a Factory while the same is in motion for the Purpose of propelling any Part of the manufacturing Machinery; and no Child or young Person shall be allowed to work between the fixed and traversing Part of any self-acting Machine while the latter is in motion by the Action of the Steam Engine, Water-wheel, or other mechanical Power.

XXI ... That every Fly-wheel directly connected with the Steam Engine or Water-wheel or other mechanical Power, whether in the Engine House or not, and every part of a Steam Engine and Water-wheel, and every Hoist or Teagle, near to which Children or young Persons pass or be employed, and all Parts of the Mill-gearing in a Factory, shall be securely fenced; and every Wheel-race not otherwise secured shall be fenced close to the Edge of the Wheel-race; and the said Protection to each Part shall not be removed while the Parts required to be fenced are in motion by the Action of the Steam Engine, Water-wheel, or other mechanical Power for any manufacturing Process.

The Act also required the notice of any accident that occurred in a factory which prevented the injured person from returning to work before nine o'clock the following morning. This had to be reported to the Surgeon appointed to grant certificates of age for the district in which the factory was situated. The Surgeon was required to send details of the accident to the Inspector. Penalties were imposed by the Act; for instance, the penalty for not fencing machinery required to be fenced was not less than £5 and not more than £20. If the failure to fence the machine, after notice from an Inspector, resulted in injury to any person, the penalty was not less than £10 and not more than £100.

With the increasingly technical nature of the legislation it was now necessary to provide an interpretation of the requirements to remove any doubts about the meaning of the words, for example:

Mill-gearing − shall be taken to comprehend every shaft, whether upright, oblique, or horizontal, and every Wheel, Drum or Pulley by which the Motion of the first moving Power is communicated to any Machine appertaining to the manufacturing Process.

Factory − shall be taken to mean all Buildings and Premises ... wherein or within which Steam, Water or any other mechanical Power shall be used to move or work any Machinery employed in preparing, manufacturing, or finishing, or in any Process incident to the Manufacture of Cotton, Wool, Hair, Silk, Flax, Hemp, Jute, or Tow, either separately or mixed together ...

With the passing of the 1844 Act, Inspectors turned their attention to enforcing the safety clauses contained in the Act. Leonard Horner directed proceedings against a company in Bolton within a month, following an accident in which a woman was scalped by an unfenced shaft. The magistrates dismissed the case and there were other cases with a similar outcome. Horner, who was responsible for the Northern District, was, much to the consternation of some owners, a great advocate of the new provisions for safety in the 1844 Act. In his report of 30 April 1845 he wrote:

> When I view the complicated machinery amongst which people work, the infinite number of wheels and other mechanisms, with projections to catch, sharp edges to cut, and vast weights to crush, the crowded state in which the machines are often packed together, and the great velocity and force with which they move, it often appears to me a marvel that accidents are not a daily occurrence in every mill.[10]

Mill owners from the cotton industries argued against the need for secure fencing of their machines. They raised many objections, particularly in respect of the need to fence overhead transmission shafts. At a meeting of Master Cotton Spinners and Manufacturers at Manchester in 1854, Mr Henry Ashworth complained that 'using wooden casing for horizontal shafts would trap cotton flue in the casing, interfere with oiling and cause fire danger'. This matter became a source of bitter conflict between the Inspectors and the manufacturers, who mustered strong political

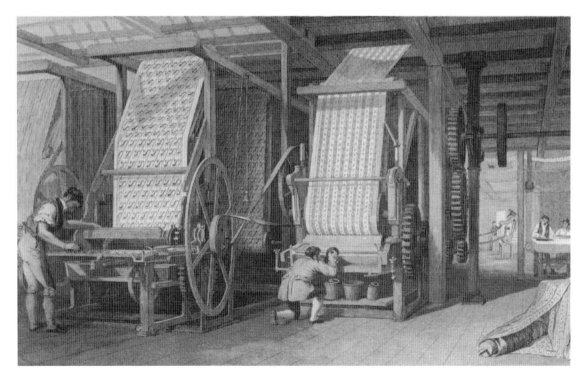

Illustration showing the dangerous parts of calico printing machinery. Drawn by T. Allom, engraved by J. Carter

Part of a self-acting mule from the 1880s, showing the many moving parts which exposed the operators to danger

support. In his report for 1854, Horner commented on the scare tactics used by the manufacturers who, from the beginning, had predicted that each new Act would lead to the downfall of British industry. In 1855, in Manchester, a group of manufacturers formed the Factory Law Amendment Association, later to become the National Association of Factory Owners, with the intention of opposing Horner and the other Inspectors, 'whose main qualification for appointment seem to have been utter ignorance of machinery and factory arrangements'. They unsuccessfully petitioned the Home Secretary to have Horner removed from his duties. Charles Dickens called this body 'The Association for Mangling of Operatives' and in his journal, *Household Words*, he wrote:

> As a consequence of this resistance, one and twenty persons have in six months been drawn into machinery and slain by every variety of torture from breaking on the wheel to being torn limb from limb. One hundred and fifty working people have had torn away from them . . . a part of the right hand that earns their bread . . . The price of life is £20; and lower damage costs but a trifle to the person whose neglect has inflicted it.

On 30 June 1856, An Act for the further Amendment of the Laws relating to Labour in Factories[11] was passed, which went some way to meet the demands of the owners. The intention of the amendments to the 1844 Act was defined as follows:

> And whereas by Section Twenty-one of the said Act (1844) it was amongst other things enacted, that all Parts of the Mill-gearing in a Factory should be securely fenced; and by Section Forty-three of the said Act Provision was made for referring to competent Persons

as Arbitrators all Questions relating to Machinery which an Inspector or Sub-Inspector might observe in a Factory not securely fenced . . . And whereas doubts have arisen as to the true Construction of said several Sections; and it is expedient that such Doubts should be removed, and that the aforesaid Provision of the said Act should be explained and amended.

The said Section Twenty-one, so far as the same refers to the Mill-gearing, shall apply only to those Parts thereof with which Children and young Persons and Women are liable to come in contact, either in passing or in their ordinary Occupation in the Factory.

The Word "Machinery" in the said Section Forty-three shall be considered as applicable to and including all other Parts of the Mill-gearing in a Factory with which Children and young Persons are not liable to come in contact in passing or in their ordinary Occupation in the Factory.

Thus the amended Act limited the extent to which machinery had to be fenced and extended the scope for arbitration in the event of differences between Inspectors and occupiers over such fencing. Inspectors were concerned about both these amendments and made their concerns apparent in their joint report of 1857 where they describe the effects of the new Act, firstly in respect of the new restrictions to the application of secure fencing:

The persons whose 'ordinary occupation' brings them near to mill gearing, and who are consequently well acquainted with the dangers to which their employment exposes them, and with the necessity of caution are protected by the law, which protection has been withdrawn from those who may be obliged in the execution of special orders, to suspend their 'ordinary occupation' and to place themselves in positions of danger, of the existence of which they are not conscious, and from which by reason of their ignorance, they are unable to protect themselves, but who on that account would appear to require the especial protection of the legislature.[96]

And secondly, they expressed concerns about the extension of arbitration between the Inspector and the occupier in the event of a disagreement between them as to the need to provide secure fencing on machinery. Arbitration was a source of some irritation to the Inspectors because the owner would appoint people who were most sympathetic to their case, usually a professional engineer or the machine maker. The joint report clearly states the Inspectors views on this matter and their preference for a legal solution:

This is in reality a question, which requires for its solution, not the opinion of professional engineers, but the evidence of intelligent and observant men who are daily employed in factories. An engineer would undoubtedly be entitled to deference in expressing an opinion whether any obstruction would arise to the action of the machinery from any particular mode of fencing, but the prevention of accidents is no part of his professional business. The secure fencing of mill gearing is therefore not a matter of opinion for the speculation of men of science, but it is a plain matter of fact, to be proved, like any other matter of fact, by evidence before a tribunal armed with all the powers necessary for eliciting the whole truth . . . In fact engineers and machine makers look only to the construction and working of the machinery, which is their business and not to the prevention of accidents, which is not their business.[96]

The 1856 Act was the first and only piece of legislation to weaken, by its amendments, previous safety reforms. It was strongly opposed by the Inspectors who were able to show in their regular reports that further more comprehensive safety legislation was needed.

The second half of the nineteenth century saw the diversification of industrial activities beyond the textile industries covered by the Factory Acts of the first half of the century. Industrial technology was moving at such a pace that the legislation was becoming out-dated and less relevant to these developments. New industries brought in their wake the need for new legislation to control the new dangers that workers were exposed to in the normal course of their employment. This period was to see the start of a programme of new Factories Acts and supporting Regulations that brought these new industrial activities into the scope of the new safety movement.

CHAPTER 2

THE CHANGING FACE OF INDUSTRY

The industrial state of the nation

During the latter half of the nineteenth century, industry underwent many changes. New developments in metallurgy and in mechanical, electrical, civil and chemical engineering technologies were being harnessed to diversify our manufacturing capability. Most of our present-day technology was in its infancy. Oil was discovered in America in 1859 and by 1875 was in plentiful supply. A four-stroke gas engine was made in Manchester in 1864 and Parsons invented the steam turbine in 1884. Coal and steam, which had been the primary means of power, were now joined by oil, gas and electricity. Machinery accidents continued to take a great toll with many fatalities reported each year, but developments in new industries such as shipbuilding and chemicals resulted in major accidents involving fires and explosions and serious property damage with heavy financial implications.

The industrial state of the nation during this period was well documented in the Chief Inspectors' reports to Her Majesty's Principal Secretary of State for the Home Office. These reports provided a valuable source of intelligence for the government of the day. As far back as 1837, the Inspectors were required to produce reports on the local state of agitation among workers in their region. For example on 3 May of that year they received a letter from the Under Secretary for the Home Department:

> Confidential. I am directed by Lord J Russell to acquaint you that he is desirous of receiving from time to time, any information which you may be able to furnish respecting the state of trade, the wages of labour and the state of tranquility or excitement in the district in which you act.[12]

This matter clearly compromised the independence of the Inspectors and came to a head in 1839 when Horner received an instruction to report weekly on Chartist agitation. This instruction was leaked by a Superintendent in Stuart's area and got into the hands of an opposition MP, who proceeded to cause the government some embarrassment. Nevertheless the reports continued to provide valuable information on the state of the industries covered by the Inspectors. The Chief Inspector's report of 1881, written by Alexander Redgrave, had the following introduction on the state of industry:

> Sir, There has been in every branch of industry increased activity, but, as a rule, the remuneration has not been commensurate with the increase of production. The operatives have been fairly well employed, and the feeling of confidence in the soundness of the gradual improvement in the condition of the country warrants the expectation that the worst has been past.[12]

The increasing diversity of trade that was taking place and now coming under the surveillance of the Factory Department was noted in the following year's report:

> Sir, At one time the state of trade in the report of an Inspector of Factories had reference only to the textile industries, but of late years every industry in the country has come under our cognisance, and a report of the state of trade must embrace a vast variety of occupations.[78]

These reports revealed that by 1880 a period of industrial gloom was coming to an end. The Superintendent Inspector for the Lancashire District, Mr Coles, reported that 4,716,700 yards of woollen cloth was exported during 1879 against 3,856,700 yards for the same period in 1878. This trend had further improved by 1882, when the Inspector reported that large stocks of cotton, which had accumulated during the depression, had been sold when the Egyptian war broke out because of fears that their cotton crop might be destroyed. By contrast the silk trade had declined and the worsted trade in Bradford continued to be unsatisfactory because of the low price of manufactured goods. The Inspector for the Midlands District, Mr Bowling, reported a steady improvement in the metal trades, noting that it was always a healthy sign when the jewellery trade experienced a revival. Silver bracelets were being turned out in Birmingham at the rate of 10,000 per week. Brass foundries were busy making chandeliers and light fittings but heavy engineering and iron founders had a bad year with only a slight demand from South Africa for the export of steam engines.

Flint glass commodities suffered from foreign competition and trade in china and earthenware from the Staffordshire potteries was no more than fair, with America being the best customer. The number of packages shipped from Liverpool between January and October 1880 was 83,357 compared with 62,362 for the previous year. Tin plate also increased in exports from £3,422,827 compared with £2,496,395 in 1879. Caution was sounded on the rapid development of the manufacture of articles in the colonies and America. This prompted the question of whether Britain was likely to regain tinned hollow ware manufacture as far as foreign demand was concerned owing to the cost of freight and duties levied in the foreign and colonial markets. The practice was to export separate stamped parts for assembly abroad. The fear was that British labour markets must suffer while foreign and colonial labour was fostered.

The rise and fall of stable industries, so common in recent years, was just as much a part of nineteenth-century economic life. HM Inspector Mostyn, for South Wales, in the 1882 report, records how several trades were at a low ebb. He reported that an increase in the consumption of steel and a decrease in iron had become more marked and the ironworks at Merthyr Tydfil belonging to Crawshay Brothers that had been idle were re-opening as steel works. On the general state of the South Wales economy he made the following observations:

> Only those who live in an iron district can judge of the dire effect produced by stoppage of one or more large works, throwing thousands of bread winners out of employ and transforming a thriving busy neighbourhood, full of life and movement into a silent dreary wilderness. My deepest impression is the quiet and uncomplaining way in which privation is endured. I hail with much satisfaction the advent of any new industry or the re-lighting of fires, which have long been extinguished.[78]

Development of the Factories Act and other Regulations

The Factory Acts of the first half of the century were no longer able to deal with the dangers that were becoming apparent in these new industries. The need to extend existing legislation was now acknowledged by Parliament. Only eleven years after the 1856 Act was passed, it was necessary to extend its scope to include other types of non-textile work activities. On 15 August 1867, an Act for the Extension of the Factories Acts[13] was passed. The main purpose of the Act was to cover new industries, the extent of which can be seen by the new interpretation of the meaning of Factory:

Factory shall mean as follows:

1. Any Blast Furnace or other Furnace or Premises in or on which the Process of Smelting or otherwise obtaining any Metal from the Ores is carried on (which Furnace or Premises are herein-after referred to as a Blast Furnace)

2. Any Copper Mill.

3. Any Mill, Forge or other Premises in or on which any Process is carried on for converting Iron into Malleable Iron, Steel, or Tin Plate, or for otherwise making or converting Steel (which mills, Forges, and other Premises are herein-after referred to as Iron Mills).

4. Iron Foundries, Copper Foundries, Brass Foundries and other Premises or Places in which the Process of Founding or Casting any metal is carried on.

5. Any Premises in which Steam, Water, or other mechanical Power is used for moving Machinery employed:

  - In the manufacture of Machinery
  - In the Manufacture of any Article of Metal not being Machinery
  - In the Manufacture of India-rubber or Gutta-percha, or Articles made wholly
    or partly of India-rubber or Gutta-percha.

6. Any Premises in which any of the following Manufactures or Processes are carried on: namely:

- Paper Manufacture
- Glass Manufacture
- Tobacco Manufacture
- Letterpress Printing
- Book-binding.

Some years later, in 1878, a new Act, the Factory and Workshop Act,[14] was passed. This brought in many new provisions for the welfare of workers, dealing with such things as sanitary provision, meal hours, holidays and certificates of fitness for employment. Sections 5 to 9 of the Act dealt with matters of safety in much greater detail than the previous Acts:

5. With respect to fencing of machinery in a factory the following provisions shall have effect:

(1) Every hoist or teagle near to which any person is liable to pass or to be employed, and every flywheel directly connected with the steam or water or other mechanical power, whether in the engine house or not, and ever part of a steam engine and water-wheel, shall be securely fenced, and:

(2) Every wheel race not otherwise secured shall be securely fenced close to the edge of the wheel race, and:

(3) Every part of the mill-gearing shall either be securely fenced or be in such a position or of such a construction as to be equally safe to every person employed in the factory as it would be if it were securely fenced, and:

(4) All fencing shall be constantly maintained in an efficient state while the parts required to be fenced are in motion or use for the purpose of any manufacturing process.

A factory, in which there is a contravention of this Section, shall be deemed not to be kept in conformity with the Act.

Where an Inspector considers that in a factory any part of the machinery of any kind moved by steam, water or other mechanical power, to which the foregoing provisions of this Act with respect to the fencing of machinery do not apply, is not securely fenced, and is so dangerous as to be likely to cause bodily injury to any person employed in the factory, the following provisions shall apply to the fencing of such machinery.

(1) The Inspector shall serve on the occupier of the factory a notice requiring him to fence the part of the machinery which the Inspector so deems to be dangerous:

(2) The occupier, within seven days after receipt of the notice may serve on the Inspector a requisition to refer the matter to arbitration and two skilled arbitrators shall be appointed, the one by the Inspector and the other by the occupier . . . The arbitrators or their umpire

shall give a decision within twenty-one days . . . if a decision is not so given, the matter shall be referred to the arbitration of an umpire to be appointed by the Judge of a County Court.

(5) Where the occupier of a factory fails to comply within a reasonable time with the requirements of this Section as to securely fence the said machinery . . . or fails constantly to maintain such fencing in an efficient state while the machinery required to be fenced is in motion for the purpose of any manufacturing process, the factory shall be deemed not to be kept in conformity with this Act.

For the first time, the Act identified dangers other than those associated with failure to securely fence machinery. Chemical and metallurgical processes were now being used in factories on a large scale and workers were coming into close contact with hot or molten liquids. The use of large grindstones, particularly in the cutlery trades, operating at high speeds was creating new hazards for workers. Both dangers were specifically identified in this Act:

(7). Where an Inspector considers that in a factory or workshop, a vat, pan or other structure, which is used in the process or handicraft . . . and near to or over which children or young persons are liable to pass or to be employed is so dangerous by reason of its being filled with hot liquid or molten metal or otherwise as to be likely to be a cause of bodily injury . . . the provisions of this Act to fencing of machinery shall apply in like manner as if they were re-enacted in this Section, with the substitution of vat, pan or other structure for machinery . . .

A typical manufacturing workshop, *c.*1869

(8). Where . . . any grindstone worked by steam, water or other mechanical p
itself so faulty or is fixed in so faulty a manner as to be likely to cause bodily inj
grinder using the same . . . the Inspector shall serve on the occupier of the factor
. . . etc.

Accidents were occurring at such a high rate that it was necessary to improve the means of recording them and where necessary to take appropriate sanctions against the occupiers of the factories and workshops. Section 31 of the 1878 Act required that any accident that caused loss of life or bodily injury should be reported to the Inspector and the Certifying Surgeon within 48 hours of its occurrence. Reporting was limited to accidents produced either by machinery moved by steam, water or other mechanical power or through a vat, pan or other structure filled with hot liquid or molten metal or other substance, or by an explosion or escape of gas, of such a nature that in the case of injury, it prevented the injured person from returning to work within 48 hours. A fine not exceeding £5 could be imposed upon an occupier if he failed to provide a written notice of the accident. Section 32 of the Act also required that on receipt of the notice of an accident, the Certifying Surgeon should conduct a full investigation as to the nature and cause of the death or injury.

> . . . The Certifying Surgeon, for the purpose only of an investigation, shall have the same
> powers as an Inspector and also shall have power to enter any room or building to which
> the person killed or injured has been removed.[78]

Enforcement of the Factories and Workshop Act

The level of fines for not complying with the Act was set at a sum not exceeding £10. If however a person was killed or injured as a result of the occupier neglecting to securely fence or maintain any type of plant or machine covered by the Act, then the fine was set at not exceeding £100, the whole or part of which could be applied for the benefit of the injured person or the family, at the discretion of the Secretary of State.

Despite the great care and attention to detail that went into the drafting of this legislation, all was not as it should have been. Annual Reports written by the Factory Inspectorate, shortly after the Act was passed show that in many respects the legislation was being disregarded or made ineffective. The following records of prosecutions provide a graphic example of this:

> 1. On 16 July 1880, James Neale a lamp maker of Birmingham was charged with not
> securely fencing every part of a steam engine in consequence thereof a certain young
> person suffered bodily injury.

> The Inspector, in his report, urged the seriousness of the offence because he had pointed
> out the want of fencing six weeks before the accident. The Magistrate expressed much
> sympathy with the defendant and he was allowed by the Bench to pay costs of ten shillings
> without a fine.[76]

2. On 17 May 1880, Thomas Pollok Hosie was charged with neglecting to maintain in an efficient state fencing of mill gearing while parts were in motion whereby Isabelle Duncan or Brown was killed on 24 April 1880.

It was undisputed that the shaft was mill gearing, and fencing was not maintained, and the shaft was within ten inches of the carding machine the deceased was cleaning. However the case was dismissed with costs of 14 shillings and 6 pence against the defendant.[76]

3. On 6 November 1880, Hirst, Proctor and Collinge, Cotton Spinners of Sowerby Bridge were charged with allowing a child to clean a machine in motion.

The child was very severely mutilated necessitating amputation of thumb and all but the little finger of her left hand. The defendant was fined one pound with 12 shillings costs.[77]

4. On 5 January 1881, Stephen Mear of High Street Saw Mills, Longton was charged with neglecting to fence the crankshaft of a steam engine, in consequence of which, James Timmis an engine driver was killed.

The accident was proved, by the finding of the body in the crankshaft pit and the shaft was not fenced at the time. The Justices held that as no one saw the accident happen it could not be proved that Timmis was killed in consequence of the neglect to fence. It was possible that the accident might have happened if the engine had been fenced, or the deceased may have contributed to it by his own act of negligence, the Justices gave the defendant the benefit of the doubt. The case was dismissed with costs of one pound and seven shillings.[77]

5. On 26 May 1881, John Hurll of Airdrie in Scotland was charged with neglecting to fence a shaft in his fireclay works, whereby a woman, Jane Davidson, was killed.

The Sheriff requested that the Home Secretary might be communicated with and favourably recommended the parents of the deceased woman for his consideration with regards to the application of the penalty. The defendant was fined £70 with one pound and three shillings costs.[77]

6. On 6 March 1882, Kneeshaw, Lupton & Co, Lime Quarries Llandulas were prosecuted for not reporting a fatal accident.

Mr Kneeshaw was Chairman of the Bench of Magistrates and retired from the Bench to plead for himself. The defendants were fined one pound with 7 shillings costs.[78]

7. On 17 April 1883, The Albion Iron Co. at Bilston were charged with neglecting to properly fence a wheel race, whereby it is supposed that a girl was killed by falling against a wheel. She was found lying outside the race with her head broken.

The Magistrate thought that under the circumstances, a mere nominal penalty would meet the case. The fine was two shillings and six pence with thirteen shillings costs.[79]

Many examples of similar injustices can be found in the Annual Reports of the Chief Inspector of Factories and Workshops. The offences taken against the employers in many cases were not for causing injury and death, but for the administrative failure to report the incident or such like. The penalties imposed by the Magistrates generally fell well short of the maximum permitted by the Factory and Workshop Act of 1878, which was £100, the whole or any part of which could be applied for the benefit of the injured person or family. It is difficult to reconcile this level of sanction with the trivial fines imposed by Magistrates throughout the country.

Inspectors regularly complained about the disregard for safety and many examples were cited in the Annual Reports of the time. The 1881 report[77] referred to young persons, who knew nothing of the cause of motion or the sources of danger that surrounded them, being put to work upon complex machinery:

> Lately a large number of gas engines have been laid down. We have had some trouble in inducing masters to fence these engines and flywheels. The drivers cannot so easily lay hold of the flywheels to give impetus to the starting engines; they fail to see that the very act, if carelessly done, is pregnant with danger. We know that the gas is not turned on full at starting time and that many drivers are not engineers, but I contend that to see a man embracing a flywheel with arms around one spoke and legs entwined around a lower one, exerting himself to the utmost to give it motion is not a work of safety.

Inspectors were quick to express their concern about the needless waste of life and limb in their reports. In their fifty-year review from 1833 to 1883[79] it was noted that in the beginning of this period there were no more than 3,094 cotton and other factories in the country. By 1883, every establishment in which mechanical power, steam, water or gas was used for the purposes of manufacture was deemed to be a factory and places where preparation of any article or adaptation for sale was carried out was a workshop. This led, year by year, to an increase in the number of accidents reported to the Factory Inspectors. They were receiving nearly 9,000 accident reports each year. This placed a great responsibility upon the Inspectors because any actions arising from their investigations needed to be such as not to hamper the process of manufacture without affording a corresponding benefit in reducing the risk of injury. The accident statistics presented in the reports confirmed their concerns. Table 2.1 summarizes the reported fatal accidents and injuries arising from machinery between 1881 and 1883.

Year		1881		1882		1883	
Group	Sex	Fatal	Total	Fatal	Total	Fatal	Total
Adult	Male	315	4027	342	4525	343	4994
	Female	6	663	11	731	1	721
Young	Male	72	1968	78	2189	47	2314
Persons	Female	10	552	5	652	4	561
Children	Male	5	288	5	307	6	303
	Female	1	101	3	97	0	100
Total		409	7599	444	8501	401	8993

TABLE 2.1: Accidents arising from machinery

The Annual Reports during this period and up to the turn of the century gave much space to the improvements taking place in the fencing of machinery. Reference was made for the first time to the provision of loose pulleys and belt shifting gear to avoid the common practice of throwing a belt off a driving wheel in order to stop the machine. The provision of belt hangers for driving belts on overhead pulleys, the improved fencing of hoists, flywheels of gas engines, circular saws for wood cutting, the intakes of crushing rollers, gear wheels on lathes, driving shafts under sewing machines and safe rests for grindstones all attracted the attention of the safety conscious Inspectorate and their determination to eliminate the needless waste of life and limb arising from accidents on these types of machine. By the end of the 1880s the factory legislation was a patchwork, brought about by the many changes and the increasing scope and complexity of the legislation since 1856.

The level of reported accidents continued to rise until the end of the century. Table 2.2 shows the total injuries and fatalities reported to the certifying surgeons between 1881 and 1896.

Year	1881	1886	1890	1892	1894	1896
Fatalities	409	386	484	425	448	596
Injuries	7150	7051	7727	8218	9301	13837
Total	7559	7437	8211	8643	9749	14433

TABLE 2.2: Accidents reported between 1881 & 1896

Consolidating legislation and the Employer's Liability Act

The next extension to the Factories Act was the Factory and Workshop Act of 1883, which incorporated white lead factories and bake-houses. This was followed by the Cotton Cloth Factories Act 1889 and shortly thereafter by an important amending Act, the Factory and Workshop Act of 1891. In addition to amending the safety requirements of the 1878 Act, the new Act had the first requirements for sanitary provisions for new premises built after the first day of January 1892 where more than 40 persons were employed. Such premises had to obtain a certificate from the sanitary authority. In addition the factory was required to provide means of escape in case of fire on all floors above ground level. Amendments to the 1878 Act had the effect of tightening the law in favour of safety. For example the phrase 'near to which any person is liable to pass or to be employed' in Sect. 5(1) of the 1878 Act was repealed and Sect. 5(3) of the 1878 Act which only covered mill-gearing was extended by inserting the words 'All dangerous parts of the machinery and every part of the mill-gearing'. But the most important innovation was the new power to lay down special rules and requirements for dangerous machines or processes:

> 8(1) Special Rules and Requirements: Where the Secretary of State certifies that in his opinion any machinery or process or particular description of manual labour used in a factory or workshop . . . is dangerous or injurious to health or dangerous to life or limb

. . . or that provision of fresh air is not sufficient or that the quantity of dust generated or inhaled in any factory or workshop is dangerous or injurious to health, the Chief Inspector may serve . . . a notice in writing either proposing such special rules or requiring the adoption of such special measures as appear to be reasonably practicable and to meet the necessities of the case . . . [15]

The Factories Act of 1891 allowed the Secretary of State to make Special Rules for dangerous machines and dangerous industries. Twenty-two Codes were made under these new provisions but only three dealt with danger to life and limb. These codes were for chemical works, explosive works and for bottled and aerated water works. The Act at Section 22(3) also contained the requirement for a coroner to be informed of a death caused by an industrial accident and to conduct an inquest into its cause. The Inspector, the occupier and any person appointed by the majority of the workpeople employed at the scene of the accident could attend the inquest and examine any witness either in person or through their legal representative, subject to the order of the coroner. The principle of arbitration, which had been in the legislation from the earliest enactment, and which had been opposed by the Inspectors over the years, was finally abandoned in the 1891 Act. Means of escape in case of fire became a requirement and there was a general widening of the scope of the regulations. This was consolidated in the 1901 Act, which now required that steam boilers should be fitted with safety valves, steam gauges and water gauges and should be examined every fourteen months by a competent person.

An important legal development of the period was the passing of the Employer's Liability Act of 1880.[16] This brought the insurance industry into the health and safety arena and gave Inspectors a valuable additional means of influencing employers to improve the standards of machinery guarding. Employers viewed the Act with some apprehension at first and large numbers joined Mutual Accident Associations to protect themselves against liabilities. The Inspectors reported that the Act was having a good effect, although there was concern about other aspects. HM Inspector Oswald of the Nottingham District, for example, reported on one case being tried under the Act concerning a man who had lost some fingers while working at a circular saw in motion. The workman was refused damages by the judge who considered the workman had been to blame in not stopping the saw. Another Inspector, Mr MacLeod, observed:

> One great evil at present exists in the manufacture of cases for the benefit of attorneys. In many instances employers as a matter of good feeling would be ready to give assistance and advantage to these who meet with accidents in their works, but as soon as legal proceedings are instituted they decline to adopt any measure of relief, which might seem to indicate a confession on their part of some culpability.[77]

The Factory and Workshop Act of 1901[17] was introduced to amend and consolidate the existing Acts. Fencing of machinery at Sect. 10 of the Act consolidated the amendments of the 1883 Act but a new safety section for steam boilers was finally incorporated into this Act as a result of the long-standing problem of boiler explosions and the practices developed by the insurance industry to overcome the problem:

11 (1). Every steam boiler used for generating steam in a factory or workshop, or in any place to which any of the provisions of this Act apply, must, whether separate or one of a range –

(a) have attached to it a proper safety valve and a proper steam gauge to show the pressure of steam and the height of water in the boiler; and:
(b) be examined thoroughly by a competent person at least once in every fourteen months.

(2) Every such boiler, safety valve, steam gauge, and water gauge must be maintained in proper condition.

(3) A report of the result of every such examination in the prescribed form, containing the prescribed particulars, shall within fourteen days be entered into or attached to the general register of the factory or workshop, and the report shall be signed by the person making the examination, and, if that person is an inspector of a boiler-inspecting company or association, by the chief engineer of the company or association.

Enforcement powers to make orders in respect of a dangerous machine were given to Inspectors in Section 17 of the Act. A court of summary jurisdiction could now, on receiving a complaint from an Inspector, prohibit the use of a machine if it was in such a condition that it could not be used without danger to life or limb, or the Inspector could prohibit the use of the machine until it was made safe to use

A boiler explosion in a brewery in 1866

by repair or alteration. Similar powers were granted in the case of a complaint from an Inspector that any place used as a factory or workshop was in such a condition that any manufacturing process or handicraft could not be carried out without danger to health or to life or limb. Another important new provision at Section 22 gave the Secretary of State powers to direct a formal investigation into the cause of any accident occurring in a factory or workshop. The Court of Inquiry had all the powers of a court of summary jurisdiction and could enter and inspect any place, or summon any persons to attend for the purpose of its investigation, and administer an oath requiring the witness to make and sign a declaration of truth of statements made in examination. On completion, the Court of Inquiry was required to make a report to the Secretary of State, stating the causes of the accident and its circumstances and adding any observations that the court thought right to make.

Factories Act legislation and its influence on the industrial development of the nineteenth century did much to improve the conditions of employment for workers in all types of industry. Reports show that Inspectors were concerned about working conditions in the brickmaking industry where girls were expected to carry as much as eleven tons of clay during the working day for earnings of *2s 3d*. The sweat-shops in East London and the dressmaking establishments of West London attracted the Inspectors' attention, not only because of working conditions but with regard to the domestic life experienced by young girls compelled by their employment to live away from home. Conditions in bake-houses were examined because of the poor hygiene and working conditions found in there. The Inspectors observed:

> The condition of bakehouses, as a matter of public discussion has been dormant for at least twenty years.[76]

The high regard for the legislation was shown by the way it was copied and included into the legislative structure of the colonial countries. In 1882 it is recorded that the Government of Bombay applied to the Factory Inspectorate for the temporary services of one of Her Majesty's Inspectors to superintend the working of the Indian Factories Act. The Inspector had to be conversant with cotton factories. Mr Meade King from the Manchester District took up this appointment. Over the years many other Inspectors followed in his footsteps in similar appointments throughout the British Empire.

CHAPTER 3

HM INSPECTORS OF FACTORIES 1833–1918

By the end of the First World War, the Factory Inspectorate had travelled a long way from its first beginnings following the 1833 Act.[5] This Act set the department in place as a new and revolutionary law enforcement authority with responsibility for the welfare, health and safety of workers in industry. When the war came to an end in 1918, Factory Act legislation and Factory Inspectors had been in existence for eighty-five years. The effect of the war upon the Factory Department, its Inspectors and supporting staff was profound because of their contribution to the war effort, both on active service and on the home front. This was the beginning of a significant period of change for the Factory Department. It was brought about by the need to adapt to the new demands of developing industry during the early years of the century, under the scrutiny of a concerned government. War and its urgent priorities required the Inspectorate to accept new responsibilities while still maintaining an acceptable level of health and safety without jeopardising the war effort. By the end of the war the Factory Department was in a better position to respond to further industrial changes that were about to take place.

The setting up of the Factory Inspectorate in 1833 followed on from the failure of the previous attempt to regulate industry by the 1802 Act.[18] The earlier Act was the product of an unreformed Parliament with its belief in laissez-faire economics, which meant that any attempt to regulate conditions of employment was looked upon with suspicion. The legislation came into place only because of the determined efforts of a few social reformers, and the legacy of a more paternalistic view of society under the Poor Law. Because of the need to provide some form of enforcement, provision was made for local Justices of the Peace to appoint two Visitors. Typically, Visitor would include a magistrate unconnected with the factory, or a clergyman. They were required to make regular visits to the mills where the pauper apprentices were employed and report to Quarter Sessions on their findings of compliance or non-compliance with the Act. The Act and subsequent amendments failed for a number of reasons. Firstly, the Visitors were unpaid and on the whole inexperienced in the work processes, often coming from the same social class as the owners. Secondly the scattered location of the mills and the inhospitable conditions that prevailed at that time did not provide a great incentive for the Visitors to apply themselves to the

task. These visits proved to be largely worthless, partly because of the
complacency on the part of the Visitors and their limited power to in
no more than £5 for any non-compliance with the Act.

Appointment of the first Inspectors

Eventually the 1802 Act was replaced in 1833 along with the formation of what
was eventually to become the Factory Department within the Home Office.
Initially four Inspectors were appointed to the post of Inspector of Factories with
responsibility for the whole of Britain and Ireland. Robert Saunders, Thomas
Howell, Leonard Horner and Robert Rickards (soon to be replaced by James Stuart)
were the four Inspectors appointed to undertake a new and revolutionary task that
was to have such a major influence in regulating industry. When they took up their
first duties in 1833 there were just over 3,000 textile mills in their charge; some
ninety years later their charge had grown to 280,000 workplaces and the number of
Inspectors had increased to 205. A salary of £1,000 was paid to the four Inspectors
out of which they had to pay their own expenses. Eight Superintendents, to be
allocated between the four districts, were employed at a salary of £250 per year. This
small band of Inspectors set about their task with great enthusiasm and dedication,
sometimes against a background of suspicion and hostility. Surprisingly, the people
most opposed to the Factory Inspectorate were factory reformers and workers. Their
suspicions were aroused by the system of patronage that decided their appointments.
All the Inspectors were active in their support of the Whigs, who had recently been
elected as the government of the day. It was taken for granted by the opponents that
Inspectors would connive with employers to make the new Act as ineffectual as the
earlier one had been. A contemporary view had this to say about the qualities of the
four men appointed:

> A briefless lawyer, a broken down merchant, a poor aristocrat and an intimate friend of
> Lieutenant Drummond . . . incompetent for their tact, but amply provided witHMost
> unconstitutional means of annoyance and mischief.[19]

The Rev. G.S. Bull, a Yorkshire clergyman who had been a leading figure in the fight
for better working conditions in factories, was equally scathing in his assessment and
had this to say:

> If these Inspectors, in whose appointment the mill owners will have due influence, should
> take the side of their patrons and masters, we shall want nothing but the torture room to
> complete their character and office as factory inquisitors.

However, the Whig Party was a reforming party and it soon became apparent that
the suspicions expressed above were unfounded. The Inspectors set about their work
in such a manner that it was not too long before the mill owners were forming into
associations to oppose some of the new requirements being forced upon them by
the Inspectors. The magnitude of the task undertaken by the four Inspectors and
their staff is recorded by Andrew Ure in his book *Philosophy of Manufactures*,[1] which

was published in 1835. This contains a statistical table, shown in Table 3.1, which lists all the textile factories in the United Kingdom at that time. The table records a total of 3,154 textile factories employing 344,623 operatives. Leonard Horner was responsible for 521 textile factories in Scotland, northern Ireland and the north of England. Saunders and Howell had 299 and 397 textile factories in their respective districts, but the greatest task by far fell upon Rickards, with responsibility for 1,937 factories extending throughout the heartlands of the cotton, wool, flax and silk industries. Detailed statistical returns such as these were to become a feature of all future annual reports from the Chief Inspectors of Factories to responsible government ministers for all factories and workshops which came within their jurisdiction. Who were these four pioneers who were able to exert such an important influence upon the industry?

Leonard Horner – first among equals

Very little is known about Saunders, although he was very committed to safety matters. Howell had a legal background and in 1822 he was appointed Judge Advocate and Judge of the Vice Admiralty Court in Gibraltar. He had no practical experience of the factory system in Britain when he took up his appointment as an Inspector. Robert Rickards was a partner in a firm of East India merchants whose appointment was approved by the *Times*:

> His experience in the world, joined to knowledge, talent and various acquirements particularly fitted him for discharging the delicate and responsible office of inspector.

Mr Rickards found the work not to his liking and retired in 1836. His position was taken over by James Stuart. Stuart had legal training but he came to the department with a controversial background. In the early 1820s the *Glasgow Sentinel* had published articles written by him on Whig ideology. Sir Alexander Boswell, a leading Tory, fiercely attacked the articles. Stuart fought and killed Boswell in a duel in 1822. He was tried for murder and acquitted in 1825, but was bankrupted in the process. In 1828, he left for America to escape financial problems. As a renowned Whig supporter he returned in 1832 to seek office in the newly elected Whig government and was appointed Assistant Commissioner for Scotland in the Factories Inquiry Commission.[12] Stuart died in 1849, thirteen years after his appointment. Only he and Horner appear in the *Dictionary of National Biography*, largely for other achievements; their work as Inspectors is only briefly mentioned.

Leonard Horner was the most influential of the four Inspectors. Born the son of a prosperous linen merchant in Edinburgh in 1785, he was educated at Edinburgh High School then later graduated from Edinburgh University where he studied geology.

Edinburgh was a great intellectual centre of the day and the home of renowned scientists and political and social philosophers. There is some evidence that Horner was an acquaintance of James Watt, the great innovator of the Steam Age. Horner was much younger than Watt, at a time when they were both members of a literary club in Edinburgh in 1805. This club, known as the Friday Club, included in its membership

Statistical Table of the Textile Factories of the United Kingdom, subject to the Factories Regulation Act, in which the Machinery is worked by Mechanical Power. Constructed from the Parliamentary Returns, by ANDREW URE, M.D., F.R.S., for his ' Philosophy of Manufactures.'

District.	Name of Inspector.	Number of Manufactures.					Moving Power.			Ages of Operatives.						Operatives in the Manufacture of								Total Operatives.	
		Cotton.	Wool and Worsted.	Flax.	Silk.	Total.	Steam Engines.	Water Wheels.	Total Horse Power.	Under 11. Male.	Fem.	From 11 to 18. Male.	Female.	Above 18. Male.	Female.	Cotton. Male.	Female.	Wool & Worsted. Male.	Female.	Flax. Male.	Female.	Silk. Male.	Female.	Male.	Female.
tland	Leonard Horner	159	90	170	6	425	224	214	10,152	285	343	6,629	14,902	8,904	19,113	10,529	22,051	1,712	1,794	3,392	10,017	185	497	15,818	40,358
rth of England	Ditto. . . .	14	24	19	0	57	21	34	895	39	34	542	1,021	685	1,261	646	1,067	290	334	336	915	0	0	1,266	2,316
stmoreland including Kendall	Robert Rickards	0	13	4	0	17	3	26	248	27	23	146	196	203	115	..	..	257	159	119	175	..	..	376	334
acashire . . .	Ditto. . . .	676	107	19	22	824	814	340	25,917	1,109	1,086	27,898	31,271	36,789	37,063	59,994	62,094	3,038	2,028	1,185	1,839	1,579	3,459	65,796	69,420
st Riding of orkshire . . .	Ditto. . . .	126	601	64	7	798	582	526	17,465	1,093	856	14,981	17,631	16,419	12,276	5,187	5,726	23,209	19,268	3,603	5,425	494	344	32,493	30,763
shire	Ditto. . . .	109	15	0	88	212	249	94	7,327	879	1,008	7,537	9,698	11,849	12,513	15,516	15,996	181	65	0	0	4,568	6,138	20,265	22,219
ptshire . . .	Ditto. . . .	5	0	0	0	5	5	4	258	..	..	202	241	250	458	452	699	0	0	0	0	0	0	452	699
byshire . . .	Ditto. . . .	60	4	..	1	65	33	63	1,436	28	28	938	1,287	2,825	2,863	2,776	2,849	35	1	..	..	30	35	2,841	2,885
ffordshire . .	Ditto. . . .	4	..	1	11	16	13	3	262	101	150	370	695	521	617	434	557	..	..	No return.		558	905	992	1,462
cestersh., &c. ‡	R. J. Saunders .	69	106	48	76	299	?†	?†	?†	1,055	1,563	2,938	6,951	3,274	8,516	3,171	5,438	1,434	2,668	404	1,085	2,258	7,847	7,267	17,030
st of England nd Wales . .	Jones Howell . .	..	319	2	25	346	?	?	?	191	193	3,646	3,009	5,017	4,905	..	..	8,234	6,564	315	377	295	1,166	..	..
th of Ireland.	Ditto. . . .	11	36	3	1	51	?	?	?	6	11	483	832	1,153	1,062	573	975	970	671	97	214	2	47	1,642	1,907
th of Ditto .	L. Horner . . .	17	0	22	0	39	17	23	895	8	13	893	2,088	970	2,650	980	1,672	..	..	891	2,479	..	..	1,871	4,151
Total . . .		1,250	1,315	352	237	3,154	?	?	?	4,811	5,308	67,203	89,822	88,859	102,812	100,258	119,124	31,360	27,369	10,336	22,526	9,969	20,438	151,079	193,544

* One-half of this power is employed in the woollen and worsted factories. + Messrs. Saunders and Howell have made no return of the steam-engines and water-wheels.
‡ Mr. Saunders inspects the following counties: Berks, Bucks, Cornwall, Derby (except the High Peak Hundred), Devon, Dorset, Essex, Hants, Herts, Kent, Leicester, Lincoln, Middlesex, Norfolk, orthampton, Nottingham, Oxford (part of), Suffolk, Surrey, Somerset (part of), Stafford (part of), Wilts (part of), Hull (neighbourhood of) in Yorkshire. There are no mills in the counties of Bedford, mbridge, Huntingdon, Rutland, and Sussex. Mr. Howell inspects the English counties which are not specified above.

Table 3.1 Statistical table of the textile factories across the UK in 1835.

many great literary and scientific people including Sir Walter Scott. One can only speculate on the thoughts of the young, liberally minded Horner when in the company of the man who had made perhaps the greatest contribution to the industrialisation that was taking place at that time. Did they, perhaps, debate the social consequences of Watt's inventive genius and did they reflect upon the awakening public conscience concerning the exploitation of the factory system that was made possible by the availability of steam? It is clear that James Watt, in his latter years, remained an impressive and respected figure. Sir Walter Scott met Watt in 1814, when Watt was seventy-eight years old. Sir Walter Scott recorded this meeting with the following words:

There were assembled about half a score of our Northern Lights . . . Amidst the company stood Mr Watt, the man whose genius discovered the means of multiplying our national resources to a degree perhaps even beyond his own stupendous powers of calculation and combination, bringing the treasures of the abyss to the summit of the earth − giving the feeble arm of man the momentum of an Afrite − commanding manufactures to arise, as the rod of the prophet produced water in the desert − affording the means of dispensing with that time and tide which wait for no man, and of sailing without that wind which defied the commands and threats of Xerxes himself. This patent commander of the elements − this abridger of time and space − this magician, whose cloudy machinery has produced a change in the world, the effects of which, extraordinary as they are, are perhaps only now beginning to be felt, was not only the most profound man of science, the most successful combiner of powers and calculation of numbers, as adapted to practical purposes, was not only one of the generally well informed, but one of the best and kindest of human beings.[27]

Horner was also a friend and supporter of Charles Darwin, who recorded in his autobiography how Horner took him to a meeting of the Royal Society in Edinburgh in 1826 when Sir Walter Scott was in the chair as President. Darwin was a student at Edinburgh University and was some years younger than Horner. hey shared a common interest in the Geological Society. Darwin moved to London in 1837 and no doubt both men continued their association. Between 1815 and 1827, Horner played a leading role in Whig politics in Edinburgh where his organisational skills were developed. A notable achievement from this period was the foundation of two educational institutions – the Edinburgh School of Arts and the Edinburgh Academy. The following account of this is given:

> In 1821, Horner had decided that mechanics and other skilled workers would benefit from instruction in those branches of science and mathematics, which would be of practical use to them in furthering their skills and abilities. At the time there was no institution to provide this. Horner's idea was that instruction was to be provided at low cost to the workmen and at hours convenient to them. The School of Art was inaugurated in the same year. Amongst the subscribers were Henry Cockburn, Francis Jeffrey, Henry Raeburn and Walter Scott. Horner was the secretary for the school for the first few years and was responsible for its administration. The School of Art was a great success. It was amongst the first of the Mechanics' Institutes. (It eventually became part of what is now the Heriot-Watt University.) Horner had a precise idea of what the objectives of the school should be. In correspondence with Sir Robert Peel when the movement to establish mechanics' Institutes had begun to grow, Horner set down his views on the content of the syllabus, given the limited amount of time the workmen could devote to study;

> "To do any real good to the workmen, his attention must be confined to such branches of science as will make him more skilful in his trade, and it will require every hour he has to bestow to enable him to acquire that accuracy of knowledge. To give lectures therefore on natural history, botany, astronomy etc. as I see announced in some of the Mechanics' Institutions appears to me worse than useless."[19]

It is clear that Horner's initiative in the establishment of the Mechanics' Institute for the education of the ordinary workman marked him out as a truly remarkable man of his time. These establishments continued to provide opportunities until recent times. Many successful engineers and technicians owed their success to the day release and evening schools and colleges, which provided study opportunities to those who had, at the same time, to continue to earn a living by regular employment. Horner's interest in this field, and his association with men of influence such as James Watt, probably explains why such an intellectual and respected individual, at the age of forty-eight, was prepared to accept an appointment as HM Inspector of Factories with the different challenges that this job presented. His interest in education continued in his new role. He played an important part in the work of the Children's Employment Commission in 1842, which considered the employment of children in mines, quarries and other trades not covered by the Factories Act. The report revealed many serious abuses and resulted in statutes controlling the employment of women and children and widened the scope of the manufacturing processes covered by the Act.

From 1827 to 1831 he was Warden of University of London (now University College, London). Shortly after this he was invited to become a member of the Factories Enquiry Commission, which was looking into the conditions of employment and hours worked by children in factories. The Commission report confirmed that children were working excessive hours and were subject to poor health and lack of a proper education. As a result of these findings the 1833 Factories Act was passed and Horner was invited to be one of the four Inspectors. Horner stands out as the most important member of the four by virtue of the way in which he applied himself to the task and because he held office for longer than any of his colleagues. He retired in 1859 at the age of seventy-four, at the end of a very controversial time for the Inspectors. For most of this time as an Inspector, he controlled the Manchester District and was at the centre of all the major safety issues in the textile trade. During this period he made himself very unpopular with the factory owners of the Manchester District. The following letter was sent in 1855 to Sir George Gray, Her Majesty's Secretary of State for the Home Department:

> We, the undersigned occupiers of factories in the district of Leonard Horner, beg to submit that the conduct of that gentleman, from his first entrance on the administration of this office to the present day, has been harsh, unfair and injudicious. It has therefore only created a feeling of distrust towards him, and increased the unpopularity of an unequal and unpopular Act of Parliament. The continuance of Mr Horner in his office being calculated to bring the law into still greater disrepute and the government into frequent unnecessary and injurious collision with the people, we earnestly solicit his removal.
> November 1855.

After his retirement, Horner continued his life-long interest in geology and devoted much of his time to the Geological Society, of which he was President between 1860 and 1862. He was a Fellow of the Royal Society and in 1861 he praised Charles Darwin's *Origin of Species* and questioned the authority for Archbishop Ussher's date for the creation of the world, 4004 BC, being in the Bible. Horner died on 5th March 1864. One of the tributes to his work appeared in Marx's *Das Kapital*:

> Leonard Horner was one of the Factory Inquiry Commissioners in 1833, and inspector of factories (substantially, censor of factories) till 1859. He rendered invaluable service to the English working class, carrying on a lifelong contest, not only against the embittered factory owners, but also against the ministers of State, to whom the number of votes given by the factory owners in the Lower House, was of more importance than was the number of hours worked by the 'hands' in the mills.[28]

Early years of factory inspection

The 1833 Act came about only after some difficulty. Despite the available evidence, owners were successful in securing the omission from the Act of any clauses concerning safety. A letter on the Factory Bill, published in 1833, had this to say about safety:

> Every practical man knows the absolute impossibility of fencing in all the machinery
> in a spinning mill which may come under the 'denomination' of dangerous. In fact the
> work could not be carried on if every part were fenced in. This is a Bill to annihilate the
> Manufacturers of Great Britain.[20]

The Royal Commission of 1833 gave careful consideration to the objections of the owners and the problems of accidents caused by machinery. Evidence was readily available that long hours of employment were a major cause of accidents. In many older mills, which were dirty, poorly ventilated and unhygienic, the lack of floor space to accommodate the new machines and transmission shafts and drives resulted in safety being compromised. This exposed children in particular to accidents from the dangerous parts of the machines, all of which were capable of inflicting horrific injuries to those unfortunate enough to come into contact with the many moving parts of the machines. Despite this evidence Parliament rejected suggestions in Lord Ashley's Bill regarding fatal accidents and his recommendation for a scheme of compensation for those who were injured, and the Act of 1833 made no provision for fencing, gave the Inspectors no power in this matter, and did not require accidents to be reported.

The concept of factory inspection by appointed civil servants was new to both the Inspectors and the mill owners, who had to submit to their enforcement of the law. There was no precedent to guide the Inspectors in standards of vigilance and propriety in their dealings with the owners. The methods of the previous Visitors provided no model for them to follow. Inspectors had to establish a code of professional conduct in relation to the Home Office and the will of Parliament whilst at the same time they had to develop an independent authority if the law was to be properly observed. Rickards wrote to his Superintendents in 1836 to remind them that they acted under the direct orders of the Secretary of State and that the code of instructions issued in response to the latter's orders must be rigidly followed. He stressed the confidential nature of their duties. They were given precise instruction on the conduct of prosecutions, their conduct towards owners, the framing of their weekly reports and what they were to look for when they visited the mills. Powers of enforcement granted to the Inspectors by the 1833 Act were far reaching and similar to that bestowed upon Justices of the Peace. They had power to make such rules, regulations and orders as necessary and these had the full force of law:

> The duties set out by the Act for the Inspectors were immense. Horner remarked "It was
> in some degree legislation in the dark". They found themselves in the roles of legislators
> and interpreters of the law as well as enforcers of the law. They had to verify the ages of
> children and young persons, effectively to be responsible for the setting up of schools and
> advise and direct the police and magistrates in the whole of Britain and Ireland – all for an
> act that was universally disliked.[12]

This level of authority added to the burden of the Inspectors and much to their relief most of it was transferred to the Secretary of State by the amending Act of 1844.[9] Great Britain was divided into four divisions for the purpose of factory inspection. Northern Ireland, northern England and Scotland formed a single division, and southern Ireland was combined with south Wales. The remainder of England was divided into South, East and Midlands Divisions. Each division had its own Inspector, each of equal status, reporting directly to the Home Secretary. Leonard Horner was responsible for the Northern Division, Robert Rickards for the Midlands, Robert Saunders for the South and East and Thomas James Howell for the Western Division. This arrangement existed for the first forty-five years until a single Department was formed in 1878. Each of the Inspectors was allowed to appoint subordinates known as Superintendents or mill-wardens. At first there were only eight of these Superintendents, making a total complement of twelve staff to inspect all the textile mills covered by the Act.

> The life of a conscientious sub-inspector had its hardships, such as those of travelling often to remotely situated mills, much argument and hostility to be overcome both from occupiers and from workpeople. In the 1836 Minute Book it is recorded that Mr Baker had broken his leg in the course of inspection and that Mr Trimmer had been mobbed by the 'factory people' in a 'country situation' near Oldham.[25]

First Inspectors of Schools

For some years the Inspectors spent much time in devising means of finding out the correct age of children and trying to regulate the hours of employment worked by the children. It was not until 1837 that the registration of births was made compulsory in England and Wales. The minimum age of employment was nine and children between the age of nine and thirteen had to attend school for twelve hours in each week. Inspectors had to enforce the requirement for part-time education of the factory children. This was the first compulsory education in the country and Horner, with his interest in education, worked hard to improve conditions. He believed that the factory children had an advantage over other children in that they received at least the rudiments of education. At the end of his career in 1859 he declared, 'In no way can the children of the operative classes be placed in more favourable circumstances than while working in a well regulated factory'. In these days the education of children was shared between Church schools, private or dame schools and factory schools. In 1846 factory schools provided 9 per cent of the teaching capacity for children. It was common practice in these schools to employ as teachers persons who had been injured in accidents and were unfit for ordinary work. Prosecution of the teachers for giving false certificates occurred and several were convicted and sent to prison for short periods. Inspectors could allocate money from fines to support worthy schools, or they could cancel a schoolmaster's certificate if he was unable to read or write. Eventually the Inspectors' responsibility for schools diminished until the Education Act of 1870 brought in compulsory education for all children and passed responsibility to HM Inspectors of Schools. In 1861, Alexander Redgrave wrote:

> One of the most useful services rendered by the Inspectors of Factories was the voluntary examination of schools. With the appointment of Inspectors of Schools the advantage of our supervision diminished. It has not been the custom in recent years to examine schools in the methodical manner formerly adopted.[96]

After 1870 the role of the Factory Inspector became the routine enforcement of the attendance at school for 'half-timers'. This responsibility lasted until 1920 when the Employment of Women, Young Persons and Children Act put an end to the practice of half-time education. The role played in pioneering compulsory education for all children is a lasting tribute to the dedication of these early Factory Inspectors.

The unrelenting machines

Despite many other responsibilities and the restrictions and limitations of the Act, the Inspectors were ever mindful of the safety issue, but first and foremost they were law enforcement officers. They were very quick to spot the dangers and their concern for the risks from 'the unrelenting machines' occupied their early reports to the Home Office. Saunders in 1835 drew attention to the serious consequences of this lack of provision for safety in the law. Rickards in one of his last reports in 1836 said, 'there was a universal cry for restriction in making powerful machines'. He observed, 'although the steam engine performed all the hard work, the machines in the hands of avarice masters was a relentless power to which the old and the young were equally bound to submit.'

These early years of the new legislation were testing times for the Inspectors and their assistants. On the question of safety, the mill owners were at best ambivalent on the need to guard their machines. Some owners suggested that the Inspectors should be empowered to direct which parts of the machines should be fenced and the owners should be relieved of all responsibility. Needless to say, the Inspectors were quick to disagree with such a transfer of responsibility. They established for the first time the legal interpretation that the owner of the machine was the person who could best foresee all the risks and apply the necessary safeguards to reduce these risks.

By 1840, experience had confirmed the deficiencies of the 1833 Act. On 3 March, Lord Ashley moved that a Select Committee be appointed to inquire into the operation of the Act and report its findings to the House. This committee confirmed the position of the four Inspectors but recommended that one of them should be appointed by warrant as Inspector General of Factories with a central office in London. He would be required to provide a written report, twice a year to the Home Office, to cover the whole field of inspection, and in consultation with his colleagues, to be responsible for drawing up rules for guidance of sub-inspectors. The Select Committee moved quickly on the other matters of safety and reported to Parliament in February 1841. The main requirements for safety were to include:

Cleaning of machines while in motion should be prohibited.
Dangerous parts of machines should be boxed in.
Upright shafts to a height of 7 feet above ground should be fenced.
Horizontal shafts within 7 feet of the floor should be totally enclosed if operatives passed beneath them.
Compensation should be paid to injured persons.

The Government was understandably cautious about accepting these recommen-dations without question. The Home Secretary, Mr Fox Maule, issued instructions to his Inspectors:

> Obtain information on those parts of machinery in mills, which without impeding the work, it is practical to box off or fence so as to diminish risk of danger to workers.
>
> Distinguish between:
> - The steam engine and water-wheels.
> - The main shaft or gearing and driving belts.
> - The manufacturing machinery, particularly to the five branches above.
>   Ascertain whether certain parts of machinery must be wiped or cleaned while they are in motion.

The Inspectors, with their interest in making progress on the safety issue, were quick to respond to their instructions. Horner in his report of 2 April 1841 believed that it was possible to frame and enforce laws that prohibited the cleaning of machines in motion. Howell on the other hand was not in agreement and a report from his Superintendent, Charles Trimner, expressed the opinion that the owners much preferred a system by which Inspectors were empowered to order fencing to be carried out if it could be done without hindrance to the work. Saunders supported Horner on the need for a new set of regulations. Deliberations during the drafting of the Bill incorporated the wishes of the Inspectors to have safety legislation. For example Section 20 of the Act of 1844 provided that no person or young person was to clean any part of mill gearing while it was in motion or to work between the fixed and traversing parts of a self-acting mule. Secure fencing was now required for every flywheel directly connected with a steam engine or water-wheel, every hoist or teagle, all parts of mill gearing and every wheel race. This protective fencing was not to be removed while the parts were in motion.

Under the 1844 Act, Inspectors no longer had the power to act as a Magistrate or to make rules and regulations. They had authority to issue notices to require the owners of dangerous machines to fit secure fencing, but the Act provided an opportunity for the owner to appeal against such a notice. Much to the dislike of the Inspectors, this appeal was dealt with by the arbitration of two examiners, one appointed by the Inspector and one by the owner. The experts thus appointed were to examine the machine alleged to be dangerous and if they disagreed with the Inspector on the grounds that the fencing was unnecessary or impracticable then the Inspector had to cancel the notice. They had authority to summons offenders and witnesses to appear in court and they had the right of entry to all the factories that came within their jurisdiction. New owners of factories were required to send written notice within one month to the Inspector, giving details about the business and the nature and amount of motive power installed in the factory.

Thomas in his study of early factory legislation[20] said that the Act of 1844 marked the opening of the third great chapter in the history of factory inspection. He noted that the authors of the Bill, because they had learned from the lessons of the past decade, were able to devise an administrative system that was to become an enduring model for all future legislation. The amending Act, apart from the arbitration by experts, was a triumph for the Inspectors and at last gave them some power to tackle a problem that had concerned them from the earliest days since 1833. Shortly after the passing of the Act, Horner directed proceedings under the new law but encountered considerable reluctance from Magistrates, who frequently dismissed cases.

Controversy over 'Safe by Position'

The fundamental debate about the fencing of machinery had now started and was to carry on for many years to come. One of the chief difficulties lay in the fact that it was often impossible for the Inspectors to decide if it was feasible to fence a machine without impeding its working. Because of the difficulty in getting a conviction from unsympathetic Magistrates, there was some reluctance by the Inspectors to insist on fencing unless there was good evidence from the accident reports from the Certifying Surgeons, and they were ever mindful of the appeal procedures that could place them in conflict with the arbitration of the experts. Inspectors gave general guidance that the in-running parts of all gear wheels should be fenced by a case of wood, iron, strong tin or strong wire covering all wheels. The mill owners were under no obligation to follow this advice. An earlier report by the Inspectors on 30 April 1846[22] reinforced the difficulties being experienced by them in interpreting and applying the provisions for fencing of machines. This matter was now proving a source of considerable embarrassment to them.

> It will be ultimately necessary to have recourse to some explanatory or amending Act of Parliament as regard these provisions of the amended Factory Act before Inspectors can enforce them satisfactorily.

Eventually the Inspectors made recommendations for further amendments to the Act in their joint report of 31 December 1846.[21]

> I. The sides of every driving drum or pulley over any strap or band shall be securely fenced to prevent straps coming into contact with the revolving shaft when it is removed.

> II. Every driving strap or band shall be provided with a guide hook or fork by which alone it shall be capable to move the strap or band from one pulley to the other.

> III. When two cog wheels of any machine, working inwards, are so exposed that any person is liable to come in contact with them, they shall be securely fenced.

The 1844 Act provided that all parts of mill gearing should be fenced, however high they were located above the ground. The factory occupiers contended that horizontal shafts higher than 7ft were 'safe by position' – an interpretation of the law that was initially accepted by the Inspectors. For nine years the provisions for the fencing of overhead shafts were enforced spasmodically until 1853, when Lord Palmerston became alerted to the apparent increasing risk of accidents caused by unguarded overhead shafting. He resolved that 'safe by position' could not be relied upon and that the provision of secure fencing in the Act must be strictly obeyed; he issued the following instruction to the Inspectors:

> That you will state for his Lordship's information what steps you have taken to prevent the recurrence of accidents in the factories of your respective districts over fencing of high shafting and gearing.

The Inspectors took appropriate action in response to this instruction and in so doing seriously upset the interests of mill owners and manufacturers. This, according to Martineau,[23] 'aroused a controversy that was to equal, if not exceed in bitterness anything which had gone before'. The Inspectors issued a circular on 31 January 1854 that stated:

> Experience convinces us that no security against accidents of the most serious kind is afforded by the position of the shaft, however elevated and that if serious accidents are to be effectually guarded against, all must be fenced. In the last three years 128 accidents have occurred from shafts so elevated as to be apparently harmless and of these accidents none have been trivial and 35 have been fatal.

Opposition to this circular was quickly mobilised by the manufacturers, and a deputation of mill owners and machine manufacturers called on Lord Palmerston to persuade him that accidents might be prevented by means other than guards. Palmerston accepted the case and directed that the Inspectors circular should be suspended. A report in the *Manchester Examiner* of 22 February 1854 conveyed the concerns of the occupiers:

> At a meeting of Master Cotton Spinners and Manufacturers at Manchester, Feb. 1854, Mr Henry Ashworth said using wooden casing for horizontal shafts would trap cotton flue into the casing and interfere with the oiling and cause fire danger.

Howell refuted this particular allegation and described it as 'the fervid imagination of the orator'. Following representation by the Inspectors to the Home Secretary, a second circular was issued on 15 March which relaxed the previous instruction by stating that for shafts positioned more than 7ft above the floor, security could be obtained by belt hangers, loose sleeving for shafts and the use of belt poles. This was followed by a third and final circular on 22 January 1885, which once more applied the word of law:

> It becomes the duty of all Inspectors to require all occupiers of factories to adopt adequate means of securely fencing their horizontal shafts.

Owners continued to oppose these requirements most vigorously and in 1855, over 700 firms met and formed the Factory Law Amendment Association, which was renamed the National Association of Factory Owners. Their main objective was

View of power loom weaving showing the layout of overhead shafting. Drawn by T. Allom, engraved by J. Tingle

to counter the requirements of the Factory Inspectors on this matter. They pushed for parliamentary representatives to bring in a Bill to get rid of the contentious legislation. They also challenged the Inspectors' qualifications for appointment by alleging their ignorance of machinery and factory arrangements. Experience strengthened the Inspectors' conviction that unguarded fencing above seven feet was dangerous. Horner's Report of 1 February 1856 noted 'he was the object of very acrimonious attitude by a section of influential mill-owners'. The mill owners and manufacturers continued with their objections and were successful in getting the law changed in 1856.[26] Horner continued in his post for a further three years to 1859 before retiring – but his vacant post was never filled.

The making of the Factory Department

The amending Act of 1844 provided the first step in establishing a unified Inspectorate. A permanent office of two rooms was set up at 15 Duke Street in London, close to the Houses of Parliament. The Inspectors requested the services of a clerk and a messenger who had to be able to write so that he could address envelopes. The clerk was a thirty-year-old civil servant from the Home Office by the name of Alexander Redgrave, who was to succeed Saunders as an Inspector in 1852, before becoming the first Chief Inspector of Factories in 1878. He resigned as Chief Inspector in 1891 after forty-seven years in the Inspectorate and fifty-seven years in public service. The office of Chief Inspector seems to have come about by a process of elimination. Leonard Horner retired in 1859 and was not replaced. Sir John Kincaid resigned in 1862, leaving two Inspectors, Redgrave and Baker. Robert Baker was a medical practitioner who was appointed as a Superintendent in 1834 and was promoted to Inspector in 1858 on the death of Howells. Baker resigned in 1878 leaving Redgrave as the sole surviving Inspector to become HM Chief Inspector of Factories and Workshops.

Alexander Redgrave was appointed as Chief Inspector at the same time as the new Factory Act of 1878 came into force. This Act was the forerunner of our modern factory legislation, covering many different types of factories and workshops. Redgrave established a structure of five Superintendents with District Inspectors allocated to five Divisions as follows:

1. Manchester Division with ten District Inspectors covering South and East Lancashire and the whole of Ireland.

2. A London Division with seven District Inspectors covering West London, South West England and South Wales.

3. A London Division with nine District Inspectors covering North East London, the East Midlands and Birmingham.

Mr Alexander Redgrave

4. Sheffield Division with eight District Inspectors covering Yorkshire, Leicester, Nottingham, Staffordshire and Stockport.

5. Glasgow Division with five District Inspectors covering all Scotland and Northern England.

The Factory and Workshop Act of 1878, which conveniently coincided with Redgrave's promotion to Chief Inspector, extended the scope of the Factory Department's responsibility. A factory was now to be interpreted in the widest sense, encompassing many new industries, each with their own particular problems of safety and welfare. Enlargement of the Inspectorate and the recruitment of new staff with some experience of these new industries was required. Understanding the problems in these industries, negotiating with different employers and employees, providing appropriate guidance on the interpretation and application of unfamiliar legislation and the enforcement of these laws were all essential elements of an Inspector's responsibility. In addition to safety and welfare, greater awareness and public concern was developing for occupational ill health problems caused by exposure to new manufacturing processes and materials.

The first Lady Inspectors

Although a large number of women were employed in the mills and factories and were shamelessly exploited, their protection had always been considered the responsibility of male Inspectors. Pressure for women Inspectors came about mainly from the Women's Trade Union Organisation and it was not until 1893 that the first two Lady Inspectors, Miss Mary Muirhead Paterson and Miss May Edith Abraham, were appointed on the recommendation of a Royal Commission. This development was not fully supported within the Department. Redgrave, in the Annual Report of 1879 wrote:

> I doubt very much whether the office of Factory Inspector is one suitable for women. It is seldom necessary to put a single question to a female employee. Possibly some details here and there might be superintended by a female inspector, but looking at what is required at the hands of an inspector, I fail to see advantages likely to arise for the ministrations in a factory – so opposite to the sphere of her good work in the hospital, the school or the home.[19]

It was the Home Secretary, Mr Asquith, who was responsible for the appointment of the first Lady Inspectors against the prophets of doom who predicted they would get their petticoats caught up in the machines. At first they worked as a separate branch under the Chief Inspector, solely on matters to do with female employment. They concentrated on the 'sweated shop' trades where female labour was exploited and on laundries where working conditions fell well below those tolerated by the Factory Act. Two of the early Inspectors, Lucy Deane and Rose Squire, had been sanitary inspectors before joining the Factory Department. They and their colleagues concentrated on matters of cleanliness, ventilation, temperature, hours of work and

Miss Mary Muirhead
Paterson

employment of young workers. They were particularly active in tackling health problems, conducting inquiries into occupational diseases such as lead poisoning, mercurial poisoning among hatters and phosphorus poisoning in match works, where women suffered from phosphorus necrosis or 'fossy jaw' as it was known. Lucy Deane was the first Inspector to raise concerns about the danger of exposure to asbestos in 1898.

At first the Lady Inspectors were not expected to deal with the safety of machinery, but in 1898 a Lady Inspector was put in charge of a district that contained a large number of laundries. This gave them the first opportunity to deal with machinery dangers. They seized this opportunity and helped to bring about the introduction of new types of guards for machines that had been the cause of many serious accidents. Records show that there was a marked reduction in the accident statistics in the following years. The Lady Inspectors were able to report that they had made up for their lack of training in engineering by consulting fully with the industry's own engineers. By the time that the consolidating Act of 1901 came into force there were six Lady Inspectors, one of whom was Miss Hilda Martindale.

The Annual Report of the Chief Inspector for 1910 records that the Lady Inspectors were investigating conditions of work in aerated water and humid textile factories, lead poisoning affecting women working in the potteries, piecework in laundries and questions of Truck. Miss Squires reported on the introduction of the Northrop automatic loom for cotton weaving, allowing a single woman to work twelve to twenty looms compared with four to six of the old style of loom.

Miss Edith Abraham

Miss Adelaide M. Anderson, Principal Lady Inspector at that time, reported on the work of her staff of seventeen inspectors, of whom six were Senior Inspectors with supervising duties. The report covered accidents in weaving factories, the most serious being due to revolving shafts driving sewing machines and lesser injuries caused by needles, resulting in sepsis. This led to a recommendation to use electro-plated needles to lessen the risk of blood poisoning. During that year the Lady Inspectors carried out inspections in 6,383 workshops and 4,482 other places that came under the Factories Act. They investigated 434 accidents of which 228 were in laundries and 206 in clothing factories. They were rigorous in the advancement of secure fencing of machinery in these premises.

This number of Lady Inspectors increased to twenty-one by 1914 when the outbreak of war brought them into hitherto unforeseen areas of responsibility. The contribution made by the women's branch during the war years was fully recognised and contributed to their eventual amalgamation with their male colleagues when the Inspectorate was reorganised in 1921. Miss Anderson wrote a special report in 1915 on the effects of the second year of the war on the industrial employment of women and girls. The following year, Miss Martindale served on a special committee appointed by the Board of Control (Liquor Traffic) to inquire into alleged drunkenness among women workers in Birmingham. In 1918, Miss Anderson CBE, who was now HM Principal Lady Inspector of Factories, was the author of a report on welfare in factories and workshops and other Lady Inspectors began to influence the work of the department. Miss Smith accompanied Mr Bellhouse the Chief Inspector of Factories and Dr Legge to an International Labour Organization conference in Washington as advisers to the United Kingdom government delegates. Miss Slocock, HM Senior Lady Inspector, was a co-author of a report on sanitation.

An investigation by the Treasury's Factory Staff Committee during 1920 recommended that the two sides of the Inspectorate, which were separately organised, should now be amalgamated and the Lady Inspector grades should be eligible for all positions within the department. Shortly after the amalgamation in 1921, Miss Constance Smith was promoted to Deputy Chief Inspector of Factories and attended an ILO conference in Geneva, which discussed general principles for the organisation of factory inspection. Recommendations were adopted by the conference based mainly on the organisation of factory inspection in the United Kingdom. Miss Hilda Martindale OBE also became a Deputy Chief Inspector and by 1932 three ladies occupied senior positions in the department:

Miss F.J. Taylor – Deputy Chief Inspector of Factories, Home Office
Miss E.J. Slocock OBE – Superintending Inspector of Factories, Southern Division, London
Miss L.M.S. Keely – Superintending Inspector of Factories, Eastern Division, Leicester

The steady progression of this branch of the Factory Inspectorate took place at a time when equal rights for women were being contested and considerable prejudice against women was a feature of the industrial and political climate of the country. It is a great credit to these Lady Inspectors that they were able to make such a major contribution, particularly to the health, safety and welfare of working women and young girls, at a time when the quest for equal rights for women had hardly begun. The suffragette movement started only in 1903 when Mrs Emmeline Pankhurst established the Women's Social and Political Union with the aim of winning women's right to vote in parliamentary elections. It took a long campaign of civil disobedience and the intervening First World War before the vote for women at the age of thirty was obtained; full equal rights for women did not come until 1928 when the age for voting was reduced to twenty-one. As for the first two Lady Inspectors, Miss Abraham became Mrs H.J. Tennant CH and Miss Anderson was honoured as Dame Adelaide Anderson. The appointment of lady inspectors increased throughout the inter-war years and by 1937 there were seventy-five women in post, all on an equal basis with their male colleagues.

The first Specialist Inspectors

In addition to machinery fencing requirements, the 1878 Act now covered industries with different safety issues. The manufacture of iron and steel, general manufacturing industries and the growth in the chemical industry needed special consideration. Some trades were inherently dangerous and were not adequately covered by the provisions of the Act. This was addressed by the Factories and Workshop Act of 1891, which permitted the Secretary of State to make Special Rules for dangerous machines and dangerous industries. This in turn required the appointment of Inspectors with specialist skills that would enable them to draft meaningful rules, capable of observance and enforcement. The Factory Department was moving, with some unease, away from the role of an administrative enforcement agency to encompass a more specialist and technical function. The method of entry into the Inspectorate at that time in common with the Civil Service was by nomination, followed by an entry examination on Latin and a modern language but no scientific subjects. Recruits were aged between twenty-four and forty years, but the applicant could be older if he came from a military service. They were expected to be in good health, with a moral character and to be free of debt.

Alexander Redgrave, when he was an Inspector in 1869, saw no need for recruits to have special qualifications, because he felt that the job consisted of the administrative duty of seeing that specified persons did as the law required. Similar views were expressed by Baker in 1858, when he stated boiler inspection should not be carried out by sub-inspectors, from which time this type of work (written into law in 1901) became the province of insurance engineering surveyors. In 1878 when Redgrave became the Chief Inspector, the thirty-nine sub-inspectors became Inspectors and from then on there was an increasing preoccupation with the technical qualifications of new recruits, determined by their entry examination. It was not until 1906 that there was any change in the entry requirements. A committee chaired by a Parliamentary Under-secretary interviewed candidates after nomination. The candidates then sat a competitive examination covering six subjects. After two years' probationary service they had to sit a non-competitive examination in factory law and sanitary science before their appointment could be confirmed.

Regulation of dangerous trades was not new to the Factory Inspectorate. Industries like pottery, lucifer matches, fustian cutting and the making of percussion caps were brought under some control by the Factory Act of 1867. When the Cotton Cloth Factories Act of 1889 came into force, it had special provisions for the artificial control of the temperature and humidity in the workshops where moisture was needed to combat excessive dryness in the cotton. This caused workers to suffer from rheumatism and chest complaints. It was not sufficient to lay down laws to deal with the problems; technical solutions had to be agreed and proved by scientific evaluation. In 1889 when the Act was introduced, Mr E.H. Osborn, an Inspector from Rochdale, was charged with its enforcement and the specialist inspection of humidity control in cotton factories. In 1899, Mr Osborn became the first Engineering Adviser. Later, in 1902, Mr G. Scott Ram was appointed as the first Electrical Inspector. In 1903, Commander Hamilton Freer-Smith was appointed as Inspector of Dangerous Trades to become the second Specialist Inspector of the Technical Branch of the Inspectorate. The Specialist Inspector Branches quickly developed from this modest beginning to carry out an important role in the ever-increasing technical complexity of industry at that time. By 1906 its size had grown to 161, comprising one third of all the Inspectors who were engineers, including a number of Whitworth scholars.

Medical practitioners had been associated with factory legislation since 1833, but only to certify the age of children presenting themselves for employment. Under the 1844 Act the Inspectors were given powers to appoint Certifying Surgeons to give 'surgical certificates of age' and under the same Act it became the duty of the Certifying Surgeon to receive and report on accidents occurring in factories. Special Rules made under the 1891 Act brought the Certifying Surgeons into the factories to carry out medical examination of workers exposed to dangerous substances such as lead. In 1896, a medical man, Dr B.A. Whitelegge was appointed Chief Inspector of Factories and two years later the first Medical Inspector, Dr Thomas Legge, was appointed to deal with ill health problems in the earthenware and china industries. Dr Legge remained in post for thirty years, during which time he investigated the ill health effects of toxic substances in the workplace. In 1910 a second Medical Inspector, Dr E.L. Collis, was appointed to investigate the problems of injury to health from dust.

By the turn of the century the increase in the level of accidents caused general disquiet in Parliament, which led to a Committee of Inquiry[24] (see Chapter 4). This Committee considered the reasons for the increase in accidents and it examined the structure and expertise of the Factory Department. It concluded that further Inspectors should be recruited with specialist experience in particular subjects and this special knowledge should be made available to other Inspectors. The accepted need to recruit new staff into the Inspectorate led to examination of the existing staff. At that time there were two hundred Inspectors occupying the following positions,

Chief Inspector	1
Deputy Chief Inspectors	2
Medical Inspectors	2
Engineering Inspector	1
Dangerous Trades Inspector	1
Assistant Inspectors of Particulars	4
Superintendent Inspectors	6
Inspectors Class I	59
Inspectors Class II	55
Assistant Inspectors	51
Lady Inspectors	18

Inspectors had a starting salary of £200 per year on recruitment, with further annual increments of £10. It was noted that a large proportion of the staff were previously trained as engineers, some chemists or analysts, some with management experience and a few with academic qualifications. It was concluded that the method of recruitment was not producing the best results, partly because of the conflicting needs for high technical ability and specialised knowledge and for the administrative skills and painstaking watchfulness required by an Inspector in carrying out his or her law enforcement duties. In considering the suitability of recruits, there was a perceived need for practical knowledge of a factory or workshop for a minimum of seven years. In addition to practical experience there was a recommended scientific training such as a Whitworth scholarship or science degree. In addition there was a recommendation to change the balance of the entry examination by substituting a technological subject for modern history.

The recommendations of this Committee endorsed the formation of specialist branches within the Factory Department. By so doing, this would in due course enhance the department's ability to respond to the rapid growth and changes in technology and the new risks to health and safety this presented. It was an opportune recommendation, which ensured that the Inspectorate would command the respect of industry during an important period in the country's industrial growth. The intervention of the First World War prevented the recommendations being acted on immediately. The model of the specialist branches set by the Medical and Electrical Inspectors did not happen in engineering until 1921 and it was 1944 before this extended to the creation of an enlarged Engineering and Chemical Branch.

CHAPTER 4

IN STEP WITH THE TWENTIETH CENTURY

Industrial development

The beginning of the twentieth century saw a continuing growth in trade and the development of new industries. The Chief Inspector of Factories' report[80] for the year 1910, to the Rt Hon. Winston Spencer Churchill, the Secretary of State for the Home Department, gave a graphic record of the rapid changes taking place at that time. Advances were noted in the use of electrical power in factories and the Inspectorate were concerned that although many companies were making use of electrical energy, they knew little or nothing about the hazards or their duties under the Electricity Regulations. Electrical accidents in factories in 1910 amounted to 276, including five fatalities. Workshops and factories, with the installation of an electrical supply, were now able to use modern machines such as automatic lathes, milling machines, woodworking machines and power presses. Power driven lifts and hoists, and mobile and overhead travelling cranes all added to the hazards of a modern workplace of the period.

Dockyard extensions and ship repairing work in the South West District was noted, with consequential growth in the production of steel bars and the tin plate and galvanising trades in South Wales. It was reported that in the Bristol Channel there were thirty-seven dry docks, four pontoons and three slipways, all well equipped. Other developments noted in the region included a fully equipped electric lighting plant, and extensive grain warehouse at Avonmouth, and a new factory for the manufacture of aircraft at Filton. Enforcement of the Docks Regulations on the Thames was difficult because of the extent to which the loading and unloading of ships was carried out. In the Port of London alone, there were 400 contravention notices issued concerning the Docks Regulations. There were 1,107 accidents, including twenty-three fatalities, notified during the year. The port employed 13,000 people to deal with the cargo from the 10,695 ships that entered the port. Two serious accidents occurred during dock construction in Newport and Birkenhead, resulting in a special Departmental Committee being set up: 'To enquire into dangers attending deep excavations in connection with the construction of docks and similar works and to consider and report what steps could be taken to minimise such dangers.'[80]

Accidents in shipyards were increasing too and new methods in rigging up gangways and erecting and securing staging were introduced to reduce the number of accidental falls from height. Access to and from the inner reaches of ships and the close proximity to crane movements created dangers, particularly for the young and the foolhardy. Irregular employment with long hours and piecework conditions led to labour troubles in the yards and were contributory factors to the poor safety record. Although a lock-out for three months on the Clydeside shipyards had affected other iron and steel trades, the output tonnage from the shipyards was still better than in 1909. Several new works opened in the West of Scotland, including a torpedo factory at Greenock.

Factory Inspectors' reports from other districts highlighted the variety of industrial development that was taking place. From the North East it was observed that Newcastle had introduced new machines for the manufacture of chains, while in the North West District, although employment was adversely affected by short-time working in the cotton trade, there was a revival in the weaving trade. Four new spinning mills were opened in Oldham and the spinning of fine worsted was introduced to Bolton by a French firm. Reports by HM Lady Inspectors indicated growth in wholesale dressmaking, millinery and workshops making high-class clothing. A rapid growth in the laundry industry was noted together with improvements in the application of automatic guards on the laundry machinery.

The Inspectors' report for the South Eastern District recorded a notable increase in the production of motor cars, and modern factories for the design and manufacture of aircraft. The increased public appetite for the cinema boosted trade in the manufacture and processing of celluloid film and the development of the cinematography entertainment business. Factories for making filament lamps, boots, shoes and clothing were flourishing. Cement works in Kent were busy too. In the Midlands, it was reported that there was an increase in the manufacture of heavy ordnance at Coventry, improvements in the pottery trade in Stafford and shoe and hosiery factories had been extended in Leicester; also a new carpet factory in Worcester had been opened.

Increasing risk of accidents

So many new developments were taking place within such a short period of time that it was almost inevitable that the accident reports would not only increase in number, but also in their complexity. Accidents in factories involving machines tended to affect no more than a few persons, but industry was now more complex and there was a higher risk of the accidents involving many workers and even members of the public. By 1910 the number of fatal accidents when compared with 1896 (see table 4.1) had doubled to 1080. This apparent increase in the accident rate was explained mainly because many more industries were now covered by the Act and the population of employed persons had increased. Table 4.1 shows this trend for the first ten years.

Year	All Accidents		Machinery Accidents		Explosions	
	Fatal	Injuries	Fatal	Injuries	Fatal	Injuries
1901	1035	29297	361	24440	50	1442
1902	1110	30076	417	25174	34	1341
1903	1047	30509	379	25400	38	1467
1904	1018	29961	247	25141	31	1097
1905	1063	32002	360	26953	48	1048
1906	1116	35696	366	30119	55	1233
1907	1179	43478	420	35148	35	1467
1908	1042	41901	363	33629	34	1307
1909	946	39966	325	32197	28	1081
1910	1080	43791	372	35524	28	1109

TABLE 4.1: Accidents reported between 1901 & 1910

The Factory Inspectorate had reason to be concerned about machinery accidents in factories. Their concern was illustrated by examples given in their reports. Accidents of the most terrible kind continued to be caused by revolving shafts. Workers' clothing or hair became entangled and drawn round the machine. It was difficult for the Inspectors to convince workers of the risks they were exposed to by this type of hazard.

Grindstone and emery wheels continued to cause accidents when the wheel fractured at speed. It was noted that the majority of the accidents occurred in the first week of running a new wheel, pointing to deficiencies in the selection and method of mounting the wheels. Mr Garvie of Wolverhampton wrote:

In one case (fatal) I calculated the speed as at 3,875 feet per minute, which was more excessive because the stone was not of a truly homogeneous nature and composition, being unequally hard and soft. The grinders declare they cannot earn big wages on a "slow wheel"; they mean a small stone, they call it slow, whereas its actual velocity is usually greater than a larger stone. A larger stone earns more money, its larger (flatter) curve exposes a larger grinding surface at any one moment, whereas the smaller diameter has necessarily a sharper curve, and the correspondingly smaller grinding surface can, of course, only execute a smaller amount of grinding, with a little reduction in piecework earnings; but the grinders put earnings before safety in trying to make a small stone grind as much work in the same time as a big stone can grind.[80]

The technical problems of securely fencing the newer more complex type of machine continued to tax the ingenuity of the Inspectors. Accidents caused by power presses, which presented a comparatively new danger, were at that time being contained because of the development of automatic guards, but efforts were still needed to improve the efficiency of the protection. Some progress was noted in the fencing of milling machines, although there remained some reluctance on the part of owners to fit guards to milling cutters:

Much progress was made with the fencing of milling cutters last year, when four successful prosecutions were taken for delaying to provide guards. The opposition appears to have

Above: Operator's hair trapped by a drilling machine

Left: A printer trapped by a platen printing press

A grinding wheel accident

subsided and the fencing of these cutters now takes its due place in the ordinary routine of our work. When cutters are dangerous by position, construction or working, guards are requisitioned, but not otherwise. One of the accidents last year caused the amputation of a boy's right arm and shoulder, and in another, the use of a man's right hand was impaired or destroyed by the tendons of his wrist being severed. Both these accidents would have been prevented by simple hoods over the cutters. This had been previously requested by us.[80]

The proper fencing of platen machines in the printing trade caused difficulties, but Mr Vaughan, Superintendent Inspector for the South Eastern District, was hopeful that an automatic feed for platen machines that he saw demonstrated at an exhibition would solve this problem. Woodworking machines attracted attention to ensure that guards were used and properly fitted. The same Inspector reported that failure of workers to properly adjust the guards on circular saws to suit the work being done was a cause of accidents. He observed that it was now rare to find a machine to which guards had not been supplied but it was equally rare to see them properly used:

> Employers think their duty ends with the provision of fencing and workmen (especially in the country) appear to think they are entitled to risk an accident if they prefer to work without a guard.[80]

Accidents in the textile industries caused by unguarded gearing, rotating shafts and flying shuttles still continued at a high rate. Lady Factory Inspectors were now playing a major part in the work of the Department. Miss Adelaide Anderson reported on the work of her staff of seventeen Lady Inspectors. A report is given of an accident to a woman who had both hands drawn into the rollers of a calender at an institution. The necessity to provide a guard had been previously pointed out but the advice had not been taken. Proceedings followed with a subsequent civil action, which resulted in heavy damages. The Inspectors, nevertheless, were satisfied with the good progress that was being made by the laundry industry in developing effective safeguards for their machines.

The Inspectors highlighted other types of accident. The severity of injuries from lift accidents continued to cause concern. It was reported that this was making occupiers recognise their responsibility and respond to the Inspector's advice. Many lifts were being fitted with electrically controlled safety apparatus. It was believed that this would lessen the possibility of accidents and remove the danger of starting the lift when people were getting in and out. One fatal hoist accident was reported as an example of negligence in failing to provide proper fencing:

> The hoist in question was used for the purpose of carrying material to the top of a blast furnace. Fencing had formerly been provided to the hoist well at the top of the furnace but had been broken. A number of men were engaged on the top of the furnace adjusting the hopper when a slight explosion occurred and one of the men in his fright turned and ran into the hoist well; the cage happened to be at the bottom and he fell a distance of 80 feet.[80]

Accidents caused by cranes and lifting equipment were increasing in number. This was attributed to the increased use of electric cranes, which were faster and more powerful. This put greater responsibility upon the drivers who, unless they were careful and well trained, could easily over wind the crane resulting in the breaking of the rope and dropping of the load. One Inspector reported of three such incidents taking place within a short period of time in the same works, one of which resulted in a fatality. Devices to prevent over winding and overloading the crane were now being asked for to prevent these types of accidents.

Other more extensive accidents and dangerous occurrences were reported. Failure of steam boilers often resulted in explosions that caused destruction on a large scale. The expanding chemical industry had its share of serious incidents. Fires with heavy loss of life, particularly in crowded multi-storied factories, were an increasingly worrying development of the period. These were well illustrated in the reports:

> A fire broke out in a Birmingham stud and button workshop. Eighteen persons, mostly young girls were employed in the workshop. Eight girls worked on the top floor. The available means of exit consisted of an internal wooden staircase from the top to the first floor and below this an outside stair to the ground. The outside stair extended to the top floor but the workshop door on this floor was padlocked on the outside. The fire broke out on the first floor when two girls were searching for a tool under the bench near a bag containing about 5 lbs. of scrap celluloid. They were using a lighted match or piece of paper when suddenly the bag burst into flames. The top of the room, which was not ceiled, was covered with paper. In about a minute the internal staircase was choked with smoke and flame. Only one of the eight girls came down it. The others were imprisoned above until the occupier went up the outside stair and burst the door in. He found four of the remaining seven girls beside the door and got them out. The other three girls had

apparently run panic-stricken to the end of the room furthest away from the door. The smoke was so thick the occupier could not get to these girls. The girls were heard breaking the windows and trying to get through the small iron frames. They were got out about ten minutes later by firemen; two of them were dead and one died on the way to hospital.[80]

Another fatal accident involving steam plant was reported by an Inspector from Birmingham. This was caused by the bursting of a cast-iron cylinder on a back washing machine in a wool spinning factory. The steam pressure was about 80psi; but the end of the cylinder, which blew out was only ⅝in thick and unsupported by ribs. The machine had been used for 20 years at a pressure of 60psi. The boiler pressure had been increased but no enquiry had been made regarding the strength of the cylinder, and the cylinder, which was a steam receiver, had not been examined by a boiler surveyor. The inspector concluded that the stresses must have been considerably beyond the safe working pressure.

Under the Boiler Explosions Act of 1882, explosions had to be reported to the Board of Trade within twenty-four hours of the occurrence. During 1910, the Board of Trade carried out three formal investigations and forty-nine preliminary inquiries under the Factory Act, which in total had caused four deaths and thirty-nine injuries. There were twelve boiler explosions involving seven water tube boilers, three Lancashire boilers and two vertical boilers. The remaining forty explosions were due to failure of other appliances − including economisers (1), evaporators (1), bakers' steam oven tubes (14), high pressure heating boilers (1), steam pipes (5), steam valves (2), boiler feed pump (1), cast iron feed pipe (1) and steam receivers (14). The report implied negligence on the part of the owner in sixteen of the incidents.

The chemical industry attracted interest and reference was made to an explosion of 20 tons of chlorate of potash at Liverpool on 21 July 1910, when three workers were injured. This set up inquiries into the precautions to be taken in works where chlorates of potash, soda or baryta were made or stored in bulk. Fifteen previous explosions, with one exception, had been due to chlorate of potash used in chemical works, match factories and explosives works. It was concluded that chlorate of soda used in aniline black dyeing and in calico printing works was probably less dangerous. Chlorate of baryta used in the making of fireworks and in calico printing had no record of explosions.

A Committee of Inquiry into the Factory Department

Between 1900 and 1907 the total number of reported accidents rose from 79,020 to 124,325. The textile trades, metal founding industries and engineering all showed substantial increases in non-fatal accidents. This led to a motion being laid in the House of Commons in 1908 concerning accidents in factories and the organisation of the Factory Department:

That this House is of the opinion that the increase in fatal and non-fatal accidents in places under the Factories and Workshops Act is of such a character as to demand immediate attention, more particularly so far as it bears upon the organisation of the Factory Department of the Home Office.

During the debate, special attention was given to the method of appointing Factory Inspectors and to the adequacy of their numbers. Winston Churchill, the Home Secretary, promised to appoint a Departmental Committee to consider the question of the increase in reported accidents. The committee, which included notable MPs such as Mr J.M. Ramsey MacDonald, took evidence on forty-one days, examined fifty-eight witnesses and visited factories. They examined representatives of insurance companies involved in workman's compensation business, made special enquiries regarding boiler explosions and invited makers of machinery to answer questions on the fitting of guards to their machines. The committee finally produced its report in 1911.[29] The findings gave a clear insight into the workings of the Department and the recommendations formed the basis of the Factory Inspectorate's policy for many years to come.

It was concluded that there had been a steady enlargement in the area of accident risk due to the gradual growth in trade over the last twenty years. A number of causes were identified, including the increased speed of machines and the pressure of work. The introduction of piecework and an increase in the level of fatigue had some effect on the accident rate. They commented that the increase in the number of machines in use could be expected to increase the accident risk. The use of electric motor and gas engines had made it easier to obtain power and new inventions enabled machinery to perform operations that could previously only be done by hand. On the question of improvement in guarding of machinery, the committee concluded that machinery was now better guarded than formerly. There was a better knowledge of the points of danger and the guards were more efficient. They were satisfied that an appreciable decrease in the accident risk had occurred as a result of this improvement. The Workman's Compensation Act had in their view been beneficial in directing employer's attention to the occurrence of accidents. They did not accept the view that safety was weakened by the practice of insurance on the grounds that this relieved the employer of personal responsibility by the payment of a premium, nor did they accept that insurance companies had little interest in reducing accidents so long as they could charge premiums that covered their costs.

Nevertheless the committee considered that the accident risk remained higher than it need be and they made a number of proposals to prevent accidents occurring at such a high rate. They proposed that Inspectors should become specialists in accident risk. Better communication between the Inspectors and industry could be achieved by organising conferences on special subjects. The committee acknowledged that the Inspectors could not be experts in all the industries they inspected, no matter how able and qualified they might be. It was important therefore that certain Inspectors should specialise in particular subjects and their acquired special knowledge should be made available to all the staff and to all occupiers concerned through the publication of reports and memoranda. The committee also recognised a concern of the occupiers, who complained about the lack of uniformity in the safety requirements of various Inspectors. In addition to the need for conferences, the committee also recommended that a Safety Museum should be set up for the benefit of setting standards in safety appliances. It was recommended that amendments should be made to the legal provisions of the 1901 Act and a greater use should be made of the powers within the Act to make special regulations on safety matters.

The accepted need to recruit the highest calibre of staff into the Inspectorate led the Committee to examine the organisation and experience of existing staff. It was the majority view of the Committee that the method of recruitment was not producing the best results, partly because of the conflicting needs for high technical ability and specialised knowledge and for the administrative skills and thoroughness required by an Inspector in carrying out his or her law enforcement duties. An interesting footnote to the work of this Committee was given in further notes on the Factory Inspectors written by the two MPs on the Committee, McDonald and Gill:

> The idea that workshop experience and high technological knowledge can be combined in the same person is undesirable. This leads to division in grades, inefficiency and waste and resentment felt by the lower grades. We believe instead of putting on the ordinary inspecting staff, highly technical men, special technological knowledge should be concentrated in a specialised department after the manner of the Medical Department and the ordinary work of the inspection should be done by men qualified by practical experience. Re-organising and strengthening the Factory Department must precede any narrowing of the margin of preventable accidents which still exist.

The committee was instrumental in laying the foundations for a modern Factory Inspectorate, in tune with the changing face of industry. In particular the committee gave special attention to the role and attributes of the Factory Inspector and the need for specialist skills. They were of the view that the best possible type of person should be appointed and the best use of staff should be made in preventing accidents. On this matter the committee were not in full agreement and a minority report was issued. Nevertheless their recommendations set in place procedures for recruitment, training and deployment of Inspectors, which provided the high calibre of Inspector that was to prove such an asset during the First World War.

The Factory Department and the First World War

The First World War brought about great changes to industry and the role of the Factory Department. The 1915 Annual Report[81] records that many Inspectors, assistants and clerks had joined the armed forces or had transferred to other government departments concerned with the war effort. Many of the remaining staff became involved in emergency war matters, not always to do with the administration of the Factories Act. Such activities as increasing the productive capacity of munitions works, facilitating the release and recruitment of industrial workers to the armed forces and the assessment of alien internees with special work skills now commanded their attention. They organised trade conferences to consider the temporary suspension of recognised trade rules and customs and, most importantly, they facilitated the employment of women in factories doing work that had been the exclusive activity of men prior to the war. At a conference of employers and operatives in the Bristol woodworking and furniture trades, the following resolution was passed:

> This conference agrees that the voluntary enlistment of men should be facilitated and the further disorganisation of the trade through the loss of men should be prevented by the temporary employment of semi-skilled labour during the war, but that, in the opinion of the conference, this problem can best be dealt with at a national conference.[81]

This period saw a further increase in the use of machinery, longer working hours and a new and greater risk to an inexperienced female workforce. Maintaining effective compliance with the various provisions of the Factories Act was a considerable problem to the Inspectors. This was compounded by the high industrial activity, particularly in areas supporting the war effort. Industrial development was diverted towards munitions of war and equipment for the forces. New engineering works of enormous size and capacity were erected and existing factories were adapted for war work and enlarged to meet the demand. The motor car industry came to a standstill to make transport wagons, shell cases, fuses and other munitions. Other industries profited from the loss of imported goods from Germany and Austria. The chemical industry manufactured high explosives and aniline dyes. Toy making, lithographic pottery transfers, beeswax refining, bronze powders for lithographic work, dry batteries, fancy leather goods and electroplating were all new. More traditional industries were stimulated, including hosiery yarn, absorbent cotton wool, weaving fancy tablecloths, flint glass and the glass bottle trade, and the brewing of light beer in place of German lager beer.

Recruitment of more Lady Inspectors had followed the introduction of women into factory work. A special report was written by Miss Anderson,[30] which examined the effect of industrial employment on women. Another Lady Inspector, Miss C. Smith, reported as follows:

> As regards English women: the revelation of their own hitherto unsuspected and undeveloped capacity has undoubtedly come as a surprise to many women engaged in unfamiliar war work. They have learnt that they are capable of better things. With training in new and more complicated operations the new element of interest has been brought into their work.[81]

In this, the third year of the war, the Factory Department was largely employed in war related work. The 1916 Annual Report[82] records that forty-five members of staff had joined the forces and a number of these had been killed or injured in action. The remaining staff worked under great pressure to secure suitable fencing of machinery, adequate means of escape from fire and good sanitary and welfare conditions. A further report on the employment of women was issued[31] and trade conferences on this topic were held with a number of industries where women were finding employment. It is recorded in the 1917 Annual Report[83] that women were now to be found working in ship and marine engineering yards, blast furnaces, forge works, copper works, spelter works, construction works, large electrical stations, maintenance in gas works and other intensive manual activities.

As the war neared its conclusion, the Chief Inspector of Factories in his report for 1918[84] was able to review the effects of the war upon industry and its workforce. The outbreak of the war had an immediate effect upon industry, caused by financial pressures and difficulties in getting raw materials. A period of depression prior to the war was replaced by demand for war equipment and articles hitherto supplied from abroad. Works and machinery had to be adapted to the demands of the war effort. This was supported by the zeal and patriotic spirit of the workers. The question of safety was always an important matter for the Inspectorate, especially during the war years with the altered conditions of employment. Increased use of new types of machinery and changes in the workforce with more women workers might have lead to an increase in the level of accidents, but this was not the case. Statistics showed a decrease; for example in the Midlands the total number of accidents in 1914 of 35,706 had fallen to 28,699 by 1918. It was noted, however, that the decrease was entirely in non-machinery type accidents. In fact during 1918 the number of fatal machinery accidents had increased. Difficulties in obtaining labour and material for the manufacture of machinery fencing probably accounted for some increase in machinery related fatalities.

This realisation brought about a new thinking on the part of the Inspectors. They concluded that the reduction in non-machinery accidents was probably due to the greater care by managers and foremen in works where women were employed and this thinking led to the Safety First Movement, which was from now on to be fully supported by the Inspectorate. Co-operation with management and workers was encouraged. Attention was given to signs and notices that pointed to the dangers and need for care. The object of Safety First was not only to ensure that guards were in good working order and regularly used, but to prevent accidents due to inexperience, lack of thought and carelessness. The formation of small safety committees in factories and self-inspection were now to be actively encouraged. A meeting of employers and workers was convened by the Lord Mayor of London to inaugurate the Safety First Movement and the British Industrial Safety First Association was formed.

The Inspectors expressed two fundamental safety principles in their report:

A great work would be accomplished, if trade-by-trade, the fencing of machinery could be standardised. The hands of inspectors would be strengthened enormously thereby, and the opposition, which they now too frequently meet would largely be removed.

Accident analysis. If accidents are to be prevented, knowledge of how and where they occur is necessary. This can only be by full and careful analysis of accident reports. It is not sufficient to know that accidents have occurred in such and such a department or on such and such a machine. It is necessary to get down to much closer detail, and to examine precisely by what portion of the machine it was caused. Accidents tabulated on these lines will at once show which are the danger zones and what are the exact dangers to be guarded against.[84]

These principles, although stated in 1918, remain valid and represent the basis of much of modern thinking on safety. Standardisation of plant and machinery and methods of safeguarding these are major contributions to accident prevention. Accident analysis leads to risk assessment and the provision of practical safeguards, commensurate with the hazard and the degree of risk associated with that hazard.

On a more sombre note the 1918 report concluded that probably few departments in the Civil Service had done more varied and useful work during the war than the Factory Department of the Home Office. Out of a male staff of 262, many over military age, 102 had enlisted in HM Forces. Fifty-six other members of staff had been seconded to other departments. Their experience and knowledge was especially useful to the military and civilian technical units to which they were assigned. Many had received promotion and had acquitted themselves well. Some still suffered from their wounds and effects of their service. Six Inspectors and five Inspector Assistants and Clerks were killed in action. Those who had served in the forces were listed in the 1918 report. The eleven members of staff who had been killed in action are listed below.

Inspectors killed in action

Name	Rank	Regiment	Date
W.J. Law	Major	7th Lancashire Fusiliers	20.12.15
C.E. Pringle	2nd Lieutenant	Loyal North Lancashire Regiment	11.07.16
D. Buchanan	Private	17th Batt. Manchester Regiment	20.07.16
A. Clark-Kennedy	Captain	5th Batt. King's Own Borderers	19.04.17
C.N. Loveday	Private	21st Batt. Royal Fusiliers	12.07.17
L.P. Evans	Lieutenant	6th Batt Royal Fusiliers	21.08.18

Inspectors Assistants killed in action.

D.G. Moore	Lance Corporal	4th Batt. King's Liverpool Regiment	06.09.16
G.E. Smith	Private	Kings Own Scottish Borderers	20.08.17

Inspectors Clerks killed in action.

A.V. Harris	Private	17th Sherwood Foresters	05.11.16
J.L. Charlton	2nd Lieutenant	Machine Gun Corps	15.09.18
A. Upham	Corporal	11th Border Regiment	10.07.17

CHAPTER 5

A SECOND INDUSTRIAL REVOLUTION

Electricity – the new source of power

The Industrial Revolution was made possible by the development of steam power and transportation in the form of the railway system and canals with good links to the ports on the coast. A second Industrial Revolution took place during the early twentieth century for similar reasons. In this case the source of power was electricity, with its convenient distribution through a public supply system, and the development of the petrol and diesel engines in motor vehicles as the means of transportation on good arterial roads with easy access to ports. Electricity as a source of power developed rapidly during the First World War. This can be seen from the increase in the number of registered factories employing electrical power and the corresponding decrease in the number of registered workshops. In 1919, for example, the number of factories had increased by 12,396 since 1914 with a corresponding decrease of 8,060 in the number of workshops. Table 5.1 shows the extent of the growth in factories and the corresponding reduction in workshops between 1902 and 1958. The first forty years was a period of great change; it took place at a time of prolonged industrial depression, which ironically was instrumental in modernising the industrial capacity of the country to an extent that matched the earlier Industrial Revolution.

By 1919, the Electricity Regulations, made under the Factories and Workshops Act 1901, accounted for 58,286 new factories that came within these Regulations. The conversion of workshops to factories by the installation of mechanical power meant that in Bristol, the supply of electrical current increased from 11 million units in 1914 to 21 million units by 1919. The use of water to generate electrical power started in Scotland in 1919, in a woollen mill on the River Tweed. The water-wheels were replaced by low- and high-pressure turbines. The low-pressure turbine pumped water through a 9in pipe to a 900ft-high reservoir during the night. During the day, the same pipe supplied the high pressure turbine. It was shown that the new system had a net gain of 355hp over the old system. Over the next five years, large power stations were built requiring electrical current to be distributed at voltages hardly thought of a few years earlier. The availability of electricity was an undeniable driving force for the economic recovery in 1931. Because electricity was so readily available, it was used extensively in the steel industry to drive the large rolling mills. It also allowed the expansion in size and efficiency of electric furnaces. There was a move towards electric drive in the textile industry, in competition with the steam turbine. From a safety point of view, this was welcome by the Factory Department as a move away from the traditional transmission drives, which had been a constant source of accidents throughout the years.

An Electricity Commissioners' Report for 1932 on the use of electricity for power purposes showed that prime movers in the order of 6.7 million horse power were still installed and powered by direct mechanical drive. It was anticipated that this capacity would eventually be transferred to the new public supply. The increased availability of this supply was made possible by the completion of the Central Electricity Generating Board national grid in 1933. The growth of the electrical industry had been remarkable, with electricity from the public supply system increasing by 14 per cent in 1934. There were changes in operational methods, most notable being local generation replaced by high voltage transmission systems. 37,719 persons were employed in distribution work in 1932, compared with 14,426 in 1924. The

Year	Factories and Workshops			Other Premises	Docks	Warehouses
	Factories (Power)	Workshops (No power)	Total			
1902	100424	145721	246145	7875	2403	4305
1909	112479	151270	263749	56607	3921	4724
1910	112370	152772	265142	77155	4051	4628
1912	117275	180802	298077	71247	4208	4555
1914	123058	178620	301678	84698	4168	4672
1919	135454	165484	300938	95340	3879	4606
1920	140064	141971	282035	130763	3913	4679
1922	137858	145684	283542	110261	3468	4642
1924	142494	133729	276223	146021	3428	4806
1926	145411	121861	267272	174798	3343	4841
1928	149532	112049	261581	224155	3244	4903
1930	154102	103371	257473	235931	3227	5037
1932	157891	90859	248749	274137	3143	4999
1936	166866	75153	243019	34615	3073	5117
1938	171825	67785	239610	41089	3125	5448
1939	173501	64669	238170	43784	3093	5527
1940	172402	61007	233409	40784	3072	5712
1941	174759	59560	234319	40853	3408	5400
1942	178454	58270	236724	39247	3150	5884
1943	181154	65364	246518	39128	3161	6067
1944	179939	48837	228776	38727	2987	5768
1945	180944	44162	225106	40094	2927	5777
1946	189913	43358	233271	43076	2902	5613
1948	204582	38787	243369	47821	2843	5641
1950	211166	29898	241064	59094	2797	5578
1952	213759	23534	237293	90726	2677	5258
1954	212909	18955	231864	106082	2617	5499
1956	211924	16309	228233	84476	2625	5735
1958	209796	13289	223085	63603	2610	5631

TABLE 5.1: Premises under the Factories Act. 1902 to 1958.

demand for electricity increased with the economic expansion of the late 1930s, leading to the building or enlargement of many new power stations.

Aftermath of the First World War

The first year after the war went smoothly for industry. Women, who had performed so magnificently in the factories during the war, gave way to the men returning from the armed forces. Cessation of hostilities meant that established factories, equipped to supply the war effort, had to be converted to meet the needs of a post-war economy. Factories in which aircraft had been made were now making furniture, motor vehicles and wooden toys. Munitions factories were making steam boilers, locomotives, tractors and milk urns and a cartridge case factory was converted to make household hollow ware, all to satisfy an unprecedented demand. Shipbuilding and ship repair facilities were essential to the war effort and quickly expanded after the war with new yards on the east coast of Scotland at Burntisland, Kinghorn and Dundee. A yard was built in Ireland at Londonderry that employed several thousand workers. Industrial unrest during the year and a series of strikes in the North East of England resulted in a shortening of the working week to forty-seven hours.

Glass factories which were built during the war because of the necessity to obtain components previously supplied from abroad, now manufactured optical and chemical glass components and electric light bulbs. A Midlands firm which made aircraft gun sights was now making lenses for cameras and scientific instruments. New automatic machinery for glass bottle making was introduced to keep pace with stiff foreign competition. The motor vehicle industry and general engineering developed quickly in some districts. Synthetic dyes and fine chemicals including aspirin and saccharin were made in the North East. In the North West, the position of dyers and calico printers who had been hit during the war were now rejuvenated by new trading opportunities. In the Midlands, brass articles now formed by hot stamping, pressing or machine moulding replaced articles which werepreviously cast. There was a growing trade in magnetos, formerly obtained from Germany, and in the manufacture of fireside hearth furniture such as kerbs and coal boxes. Perambulators were now in demand and one firm was reputed to have 85 per cent of its work done by disabled servicemen. In this changing industrial scene, with its high demands and a shortage of skilled labour, the young and inexperienced were at greatest risk:

> In certain cases the young person had only been employed for a few weeks or less and was naturally ignorant and inexperienced. The excuse in such cases that the worker was careless is irrelevant, as indiscretion at that age must be assumed as natural, and, moreover it is seldom proved that there has been wilful carelessness. For a worker to be handicapped at the outset of industrial life be it the loss of a limb is not only a very serious matter for the victim, but a loss to the community, and even if there was carelessness the punishment is out of all proportion to the offence committed.[85]

The early industrial optimism of 1919 was quickly followed by a deepening trade recession. By the following year there were few industries not working short time and many had closed down. Manufacturers blamed this on the high cost of production due to reduced output and high wages. Labour became unsettled and there were many more strikes. This led to a lack of confidence and an inability of manufacturers to quote firm prices or guarantee delivery. An unfavourable rate of exchange and increasing foreign competition added to the problems. Small firms like laundries had to lay off their workers and alternative employment was not readily available. There were fewer opportunities for young persons to gain apprenticeships. Many workshop industries, particularly in rural areas disappeared. Employment prospects, where they existed, passed to large wholesale manufacturers and workers were forced to migrate to towns or emigrate abroad. The Superintending Inspector for Scotland observed:

> This constant migration from the rural to the urban districts, coupled with the flow of emigration to the Colonies is draining the rural parts of Scotland most seriously, and unfortunately it is usually the young, enterprising and more robust who are forced to go.[86]

The effects of the industrial depression were felt first in those districts that depended upon the traditional staple industries. In South Wales, half the tinplate, sheet metal and spelter works had closed by the end of the year. A flourishing tin industry in Cornwall had suddenly hit a low ebb. In the traditional textile trades the recession had the greatest and longest lasting impact. In Lancashire and Yorkshire the cotton trade was in its lowest condition since the American Civil War. The following observation was made in the 1920 Annual Report:

> Some conception of the extent of the enormous losses may be obtained when realised that American spot cotton which was 4.54 pence per pound in December 1914 and stood at 32.41 pence per pound in February 1920 was under 10 pence per pound on 31 December 1920. Cotton yarn, which was 12 shillings and 10 pence in February only fetched 3 shillings on 28 December.[86]

This and other problems that beset the textile industries and the cotton trade in particular persisted throughout the next two decades. It had been the growth in the textile industry that fuelled the first Industrial Revolution by creating the wealth and confidence necessary to release that essential creative drive to sustain it. The industry had long since lost these essential ingredients and was now in decline. Other industries and new technological developments were from now on to provide that driving force.

Decline of the textile industry

By 1923 the textile industry was still in a depressed state with cotton mills not working much more than half time. A similar state existed in the weaving factories due to the high price of raw materials. The British Cotton Growing Association set about developing cotton cultivation within the British Empire in Nigeria, Uganda, South Africa and Australia. Steps were taken to improve the quality of Indian cotton to make it suitable for the Lancashire mills. The raw silk industries in Macclesfield and Congleton were in a poor state owing to foreign competition and the competition from artificial silk, which was being made at Coventry and Wigan.

The flax trade in both damask and heavy linen, and the lace trade in Nottingham were similarly depressed. In the woollen trade, 40 per cent of the looms were idle in Bradford. A Yorkshire manufacturer was reported to have said 'You can't carry on trade when wool is dearer than tops, tops are dearer than yarn, and yarn is dearer than cloth'.[89] By contrast conditions were moderate in the hosiery trade and by the year 1924, the carpet trade in Kidderminster was busy along with the artificial silk industry, which continued its expansion.

The general uncertainty about the economy continued into 1925 with textiles, potteries, iron and steel and shipbuilding all suffering from acute depression. Factories for the manufacture of American cotton were on short time while the manufacture of sewing cotton was busy. Two new spinning mills were built in Oldham, the first since 1914. There was a 60 per cent falling off in worsted yarn in Bradford, where only 50 per cent of the looms were working full time. This lack of business caused anxiety in the supply of trained labour and the trade resorted to employ young persons as probationers paying them 20s to 25s weekly in anticipation of a quick recovery in trade. By 1928 the depression in trade was mainly affecting the textile and heavy industries in the Midlands, North England and Scotland. Changes in lifestyle and fashions of the population were felt by the industry. The increasing purchase of ready-made garments was the death knell of small bespoke tailors, dressmakers and milliners.

Jute bag manufacturers complained of the competition from paper containers forcing the industry to make efforts to reduce costs by making the spinning machinery more automatic. In one works the workers were able to double the amount of yarn they could spin on the older machines. Deterioration in the economic situation became acute by 1930, with the traditional industries such as textiles being the most seriously affected. Its effects were now being felt in the South of England, which had not been so severely troubled in earlier years. There was no let up in the depression until September 1931 when the United Kingdom departed from the Gold Standard. This had an immediate effect on the textile industries and there was an upward movement in the woollen and worsted trades, but not so for cotton. Full employment and shift work became possible in spinning yarn for the hosiery trade and in the weaving of lighter tweeds that had previously been imported. Other consumer-led industries also benefited such as leather, gloves, glass bottles, table glass, nails, locks, hinges, bobbins and paper.

The expansion in production following the abandonment of the Gold Standard in 1931 was short lived and by the following year, there was a general relapse to previous levels of recession. In Lancashire, the cotton and allied trades showed little sign of recovery. Cotton mills were not easily adapted for other purposes. By 1933 the woollen trade showed renewed activity and there was a marked improvement in the clothing trade in the Leeds area. The introduction of new machines and the adaptation of older machines was an important factor. Labour costs and time were being saved by the increased use of special purpose machines for buttoning, button holes, edge basting, padding, bias binding and pressing. This revival accentuated the shortage of trained workers in the clothing trade and retired skilled people were encouraged to return to the factories. Women who had been out of the trade for some years were now finding work.

Despite the general economic improvement during 1934, there were moves in Lancashire and Yorkshire to convert the redundant cotton mills and weaving sheds to other more productive purposes such as the manufacture of domestic wash boilers, printing works, ham boiling, baking and the making of electrical fittings. The depressed condition of the cotton industry had been so long known that it was almost regarded as permanent. Although the volume of trade in cotton was low, it was still regarded as a very important industry. In 1935, extensions were made to the artificial silk trade in Flintshire and a large new works was built at Preston, which provided employment for workers made redundant by the closing of the cotton mills. The more progressive employers in the cotton trade, finding little use for their old Lancashire looms, were replacing them with modern machinery, more suitable for weaving artificial silk. The woollen and worsted factories in Yorkshire were in better shape and the wholesale clothing trade in Leeds continued to develop with the decline in the retail bespoke tailoring trades. The trade introduced a conveyor system under which a garment was passed to successive workers, each of whom performed one small operation upon it. This new system concerned the Factory Inspectors, who set up an inquiry into the speed of the system to avoid fatigue and overstrain of the workers.

By 1936, the widespread industrial revival was well established in all parts of the country, only to turn again in 1937 because of the uncertain international situation prior to the outbreak of the Second World War. Cotton and woollen trades, which had shown some signs of recovery, fell back because of the difficulty of obtaining orders from abroad. The Macclesfield silk trade was busy and the demand for artificial silk and rayon was good. The Kidderminster and Axminster carpet weaving industry was particularly busy. The cotton industry in Lancashire continued its downward spiral with the closure of many mills that were turned over to the Cotton Spindles Board for scrapping. Bleaching, dyeing and finishing of cotton goods had, in consequence, another bad year. In contrast, the Lancashire woollen trade had a busy year in 1938, with an extension in the use of automatic looms. One mill was now being used to build up a mail order business for wool for knitting and rug making. This protracted decline in the textile industry was clearly recorded in the reports, which showed the number of registered textile factories and workshops fell from 11,178 in 1920 to 6,646 in 1944.

Iron and steel and the building of ships

Iron and steel production and the shipbuilding industry were, in addition to textiles, an important part of the country's economic structure. These industries, which had originated in this country during the Industrial Revolution, were to be found now in the colonies and other parts of the world. Survival was no longer guaranteed, but depended upon the ability to be competitive in a global market. The industrial unrest that struck the shipbuilding industry in 1919 had an effect upon the iron and steel industry. Surplus stocks, manufactured at inflated prices and a falling value below cost coupled with cancelled orders meant that goods had to be disposed at great financial loss. Nevertheless there were some bright spots. In South Wales a new blast furnace for 3,000 tons of pig iron a week started up. A new steel rolling mill and a new cogging mill were built in Sheffield and an extension made to the steel making plant in Stockton. Further extensions to the shipbuilding capacity took place in the North East of Scotland and the South West of England. On the River Tees, several large berths and a new graving dock were constructed. Goole on the River Humber had a good trade in the building of trawlers. Five new shipbuilding yards started up in the Newcastle area and similar developments took place along the banks of the River Clyde. Scotland saw the building of the largest iron foundry of its kind at Govan in Glasgow and several new foundries were built near Halifax.

Despite the general depression, which had set in by 1921, the investment in new and more economic iron and steel production plant continued. A new gas-producing plant at a steelworks in Swansea had an improved output equal to four ordinary gas producers with a much reduced coal consumption and lower labour costs. A new development in Sheffield saw the introduction of stainless steel as an important commodity, 'providing untarnishable metal of a ductile nature which can be worked cold into intricate shapes for purposes the earlier stainless steel was unsuitable owing to hardness'.[87] The following year saw a turn in the fortunes of the industry. In South Wales, 80 per cent of the tin plate mills were at work compared with 68 per cent at the end of 1921. Twelve blast furnaces were operating in Scunthorpe, compared with two previously. In Cleveland, the number of furnaces increased from twenty-two to thirty-six and similar improvements took place in Scotland in support of improving conditions in shipbuilding and the railways. The preceding years of slackness had enabled the employers to modernise their plant in anticipation of the improvements in trade. It was now possible to operate a 160-ton steel furnace with three operators, and a steel rolling mill was capable of producing 1,500 tons of steel rails in twenty-four hours with four operators on each shift.

The shipbuilding industry in 1923 was seriously affected by a long strike by boilermakers and output dropped to its lowest in thirty years. This had a serious effect on the many other ancillary trades that relied on the shipyards for their business. The situation improved in 1924, particularly on Clydeside, where the tonnage launched was three times greater than the previous year because of the strikes and suspended contracts. Despite this, the iron and steel trade remained below normal and many blast furnaces had by now closed down. World production of pig iron in 1925 had increased but the output in Britain had fallen by 15 per cent and Britain now occupied fourth place as an iron producer. Shipbuilding, which by its nature was subject to extreme fluctuations, was more promising during the year. The Cunard Shipping Co. order for a large passenger vessel, placed with John Brown Shipyard on the River Clyde, stimulated ancillary trades, not only in the manufacture of steel plates and angles, but paint manufacturers, upholsterers, cabinet makers, linoleum and carpet makers, and suppliers of electric light fittings, bed and table linen, cutlery, glass and many more, including marine engine builders, benefited from the orders placed by the shipbuilder.

The severe depression up to 1931 and prior to the departure from the Gold Standard crippled the heavy industries. For three successive months, only one blast furnace operated in Scotland out of eighty-three, and an average of five furnaces operated over the year. In Middlesborough, seventeen furnaces operated compared with twenty-five in 1930 and forty-three in 1929. Production of steel fell by 500,000 tons or 30 per cent of the industry's capacity, and tonnage launched on the Clyde was 70 per cent less than in 1930 which was also a bad year. On the Tyne, Wear and Tees, many of the yards had not a single ship on their stocks. Marine engineering works were not in such a perilous state because of contracts for the repair or refitting of old ships. There was no let-up in 1932, when shipbuilding saw the worst year on record. Even fishing suffered, with Yarmouth and Lowestoft experiencing a worse herring season than 1931.

Undeterred, the iron and steel industry continued to adapt to modern production methods. Old blast furnaces were replaced by more modern designs. A large iron and steel works in the North of England which had been closed since 1930 reopened in 1933, providing employment for 3,000 workers. In Sheffield, a new Siemens steel-making plant was completed with a capacity to handle 300-ton ingots. Corby, in Northamptonshire became the site for new works to utilise the local iron ore resources. Shipbuilding also adapted to modern production methods, they became more mechanised with mechanical appliances, multiple drilling machines and overhead electric cranes to lift units into place, all replacing hand bogies, punching machines and temporary rigged derricks. Electric arc welding became an acceptable method of construction. In April 1933, an all-welded ship was launched in the River Tyne. The entire hull with its frames, beams, shell plates, deck plates and bulkheads was welded throughout. It was a lighter, more rigid structure that was made at no extra cost. The Newport Bridge was built by a local firm around the same time, and was unique as it was made entirely of electrically welded joints. The increasing importance of electric arc welding in engineering construction led to the opening of a new factory in Scotland in 1935 for the manufacture of welding electrodes.

The economic recovery started in 1934 and its effects were felt by the shipbuilding industry. On the River Clyde, sixty-seven vessels with a tonnage of 268,120, including the liner *Queen Mary*, were launched compared to thirty-one with a tonnage of 56,368 in 1933. The improvement benefited the marine engineering works, where the output from one Clydeside factory alone rose from 174,674hp to 713,523hp. In the steel industry, most blast furnaces were now working and new plant was laid down to increase the output of the furnaces. Twenty-three furnaces were in operation in Middlesbrough compared with eighteen previously. More furnaces were put on blast in Northamptonshire and the new steel works at Corby was nearing completion. Here the most modern ore-crushing, grading and handling plant was commissioned to operate along with new coal-washing plant and coke ovens to supply coke for the blast furnaces. Ancillary by-product plants were installed to produce benzole and sulphate of ammonia. These developments at Corby were a necessary preliminary to the new steel works, which included a Bessemer plant with rolling mills and tube making plant. 2,500 men were needed in the production of steel tubes from ore smelted in the blast furnaces. Corby was being converted from a simple country village into a highly industrialised area with the erection of over 800 homes to accommodate the influx of a new workforce. The determination of the industry to adapt to changes in modern technology was demonstrated by their use of the selenium cell, which was a scientific novelty only a few years before. The cell in conjunction with a beam of light was used in a tube mill for the control of the automatic mechanism at the end of a roller conveyor, which by means of mechanical arms lifted the hot tubes and discharged them on to the cooling stacks. This replaced the work practice where the arms were manually controlled and accidents happened occasionally because of bad timing by the operator.

At Barrow in Furness, a new acid Bessemer plant was installed with two converters capable of producing 7,000 tons of steel per week. This replaced a smaller hydraulically operated plant with a 4,000-ton capacity yet required twenty men fewer on shift. The substitution of mechanical means for doing work of the kind previously done by unskilled or semi-skilled labour was a feature of the modernisation of the industry. In the South, the most modern blast furnace in the country, producing 700 tons of pig iron a day, was started up in Essex. In addition there was a large coke oven plant, amonium sulphate and benzole plants and an electrical generating station equipped with boilers intended to burn London refuse as an alternative fuel. The recovery of the heavy industries continued into 1935, when the output of steel was a record, due in part to many furnaces being reconstructed to improve their thermal efficiency. The year also saw a large number of men and their families transferred from Scotland to take up employment in the new iron and steel works in Corby.

Revival of shipbuilding and marine engineering was maintained and the output from the shipyards on the Rivers Tyne, Wear and Tees showed an increase of 50 per cent for 1935. Steelworks were fully occupied by 1936 and production was not always sufficient to meet the demands of the shipyards. The contrast between the activity in the Clyde shipyards and the silence of three years before was very striking. Orders for armaments accounted for some of the activity but the level of commercial engineering work was still very high. Iron and steel production reached record levels in 1937. At one of the largest plants, employing 3,000 workers, a new blast furnace, the largest in Europe, a new tar distillation plant and an additional large high frequency furnace were brought into operation. The iron and steel industry as a whole had spent £20 million in new plant during the last two years. This level of investment continued into 1937 where in one district, extensions costing £4 million included two new blast furnaces with 3,700 tons capacity a week and a battery of sixty-eight coke ovens. This blast furnace development was accompanied by improvements in by-product production plant, in coke oven design and in new mechanical handling methods. Conveyors were used to carry the coke from the ovens to the furnaces. Molten metal was poured into chills on a continuous chain instead of being cast in sand beds. Rolling mill plant had been improved and a continuous strip mill in South Wales was capable of rolling strip up to 50in wide with greatly accelerated output.

In shipbuilding, the output from the yards and marine engine works was satisfactory during 1938, although new orders for merchant tonnage were scarce. The total tonnage under construction was lower than for 1937. Orders moved from merchant ships to naval construction and in one naval yard, the number employed increased to over 10,000 workers. The docks were very busy, with Liverpool working night and day. Preston received increased imports of wood pulp, timber and petrol. Elsewhere, a dry dock was extended and oil storage tanks constructed to enable repairs to be carried out on oil tankers.

New developments – new industries

The years immediately following the end of the First World War saw the development of a wide range of new industries to cater for the needs of the post -war economy. In

1921, a large plate glass factory was opened near Doncaster and a match factory was opened in Garston in Liverpool. The chemical industry was augmented by a factory for refining edible oil and another factory in Bradford for the production of formic acid, hydrogen peroxide and lanolin from grease that was extracted from wool in the wool combing works. A new cement factory was opened in Hull in the same year. In a large number of districts, factories were opening up to make 'wireless listening sets' for the broadcasts being transmitted from the Marconi research establishments. Another important development of 1922 was the opening of a large oil refinery in Swansea that could refine 6,000 tons of oil a month. It was expected that this installation would eventually supply one third of the total petrol requirements of the country as well as other oil products.

In Sheffield, the cutlery trade was working little more than half time. Cheaper cutlery was affected by foreign competition, but there was a fair demand for the better class of cutlery. There was a demand for saws made from good tool steel. Engineering works were stimulated by export orders for railway developments in the colonies and China. In the Midlands, there were interesting developments in a new process of metal coating by a finely divided spray of molten metal. Stainless steel and 'rustless iron' had made rapid progress as a useful material. The former was now being used in connection with pneumatic machinery, stove grate work, golf club heads, shop fittings and turbines. Rustless iron, or steel, was increasingly used in the making of water taps, bath fittings, cabin fittings, finger plates for doors and other articles for which brass had hitherto been used. New materials were successfully developed for the making of mining tools that up until then had been made in America.

Trade conditions during 1924 were depressed and the number of unemployed in the country was considerably above one million; a large proportion of these were factory workers. There was concern that so many young people had no opportunity to learn a trade, which would eventually lead to a shortage of skilled labour. However, the chemical industry started to make progress. A new plant was started for the manufacture of sulphuric acid by the Schmiedal Process, which had a capacity for 1,000 tons of concentrated acid per month. Another plant was completed for the manufacture of synthetic ammonia and the largest alcohol factory in Europe was opened in Hull. As a result of experience during the war in the production, storage and transportation of liquid chlorine in large quantities, the use of chlorine increased in new processes or as a substitute for other chemicals such as bleaching powder. Another large oil refinery was opened in Grangemouth in the East of Scotland and was operating at full capacity in refining crude oil imported from Iran. It was reported[90] that ten 8,000-ton storage tanks were barely sufficient to meet the storage requirements.

Another new process for the manufacture of Endurite was started in a paper mill near Edinburgh. This used up waste products from esparto grass and other course fibres that were unsuitable for paper making. Paper mills paid large sums for the removal of this waste and it was now being collected, ground and pressed in steam and hydraulic presses to make a material that could be stained to any colour and used for combs and toothbrushes. Its principal and most important use was in the electrical industry because of its high electrical resistance. The discovery that the addition of ammonia to rubber latex, when collected at plantations, prevented coagulation enabled the material to be suitable for export in bulk. This led to further developments and the material was now being used for the production of very fine sheet rubber.

The motor industry expanded in the Midlands, in the South East of England and the outskirts of London. In 1925, the Factory Department had 14,600 works registered as being involved in this industry with more than half being repair and servicing garages. A new type of stainless steel was being produced which, while retaining the maximum corrosive resistance properties, had sufficient ductility for hot and cold pressing, could be welded, soldered or riveted and was now being used in the production of cold drawn tubes and wire and cold and hot pressed hollow ware. For some years, Sheffield had supplied to America, steel for the making of razor blades but efforts were now made to make the blades here. One firm had developed a mass production method employing 1,300 workers, mostly women, to produce one million blades a year. The introduction of internal grinding machines with automatic sizing devices, new machines for fine grinding and improved methods of case hardening were of considerable importance to the engineering industry. One firm developed methods of depositing and interlocking non-corrosive metal onto steel components. Parts of aero engines were now being treated with this process to give extended hours of service. Meanwhile the more traditional industries such as the saddlery trade were turning to other branches of the leather industry with developments in the manufacture of footballs, golf and other sports bags, tennis balls and motor cycle saddles. In the North East, a brick works with a capacity to produce 5 million bricks a year was now in operation. In the same area the development of electric light bulbs, making quartz glass apparatus for wireless installations and the manufacture of astronomical lenses took place. In the North West and elsewhere, new works opened for the production of tarmacadam to support the road-building programme. In the Midlands, in Coventry, there were plans for enlarging works with increased demands on both gas and electrical power stations to work to the limit of their capacity.

By 1928 the depression in trade was mainly to be felt in the north and in the traditional heavy industries. Many luxury trades flourished despite the general downturn in the economy, with a great demand for motor cars, tobacco, gramophones, wireless equipment and building equipment. Changes in the lifestyle of the population and its fashions had the effect of upsetting the balance of some trades. Bedsteads that used to be made from metal were now made from wood and window casements, which used to be made from wood, were now made from metal. New machines for use in the woodworking industry were welcome. Automatic machines for planing, moulding, grooving and turning were much safer to use. A new factory was equipped to make doors on mass production in order to compete with imported products. New and fashionable materials like Bakelite displaced the more traditional materials. Rooms in modern homes were now smaller requiring new designs of furniture. The desire to save on domestic staff and housework was partly responsible for the introduction of new commodities, which avoided the need for polishing. Activities in the building trade reflected in a demand for cement, bricks and tiles. Two cement factories in Kent that had been closed were re-opened to meet this demand. It was reported[99] that in brickyards in the eastern counties, it was common to see hot bricks being loaded from the kilns on to the contractor's lorries because there was no time to allow them to cool. On occasions the hot bricks burnt the bottoms of the lorries.

The general reduction in spending power caused by the high unemployment and the economic crisis of 1931 did not affect those trades involved in the luxury market. Employment in the Birmingham trades of motors, cycles and accessories had increased. In 1932 a motor firm employing 8,000 workers had increased to 15,000 workers by 1933. In a large motor works in Bedford the workforce had increased from 2,000 to 7,000. Workers travelled there from all over Bedfordshire and from London. Many workers from Scotland, Wales and the mining areas of Durham found new employment opportunities there. An industry established in the North of England made solid carbon dioxide to serve the aerated water, ice cream and refrigeration industries. An engineering works in the Midlands was already using carbon dioxide for cooling bushes and spindles to enable these to be easily fitted into holes instead of driving them in.

The cinematograph film industry was doing well with several new film studios opening in and around London, including Pinewood, which was equipped with modern editing, cutting, projection and store rooms. Trade recovery by 1937 affected industry in many fields, including the manufacture of aircraft, munitions, structural and electrical engineering, tool making, heavy chemicals, paper, cement, rubber and leather. Many small firms such as chain makers benefited. The docks also prospered, particularly in the shipment of timber. Extensions and reconstructions were commissioned on the Thames, Humber and Mersey. New industries started up including the manufacture of white lead by a precipitation process in Kent, activated carbon in Essex, magnesium and aluminium in South Wales, cellophane in Somerset, chrome magnesite bricks in the North Midlands and mass production of clocks in Surrey. The pottery industry moved towards more efficient and economic firing methods. The number of gas-fired continuous ovens in Stoke on Trent increased from three in 1932 to twenty one by 1937 and a further eleven were partly completed by the end of the year. This also led to an improvement in the health conditions in the potteries by the substitution of alumina for bedding flint. The motor vehicle and component industries remained busy. Improved methods of assembly had been perfected using new and larger power presses. Flash welding of panels reduced the use of lead filling and the associated hazard of lead poisoning. Another development was the use of broaching machines for operations previously done by planing, boring and milling machines. One such machine, which faced up engine cylinder blocks, cost £7,500 and completed in one stroke, an operation that would require several operations on other machines.

The decline in trade first detected in 1937 was partly relieved by the end of 1938 by the increasing demands of defence procurement. New aerodromes sprung up all over the country and the naval dockyards were busy. Production of guns, munitions, transport vehicles and aircraft vastly increased. It was noted[102] how the activities in aircraft construction had a far reaching effect on industry. This included the construction of huge aircraft hangars and engineering works for the manufacture and assembly of aircraft and engines. Modern aircraft were complicated structures containing over a 250,000 parts made from aluminium and high-duty alloys and containing scientific instruments, complex control systems and many other parts. The new aircraft factories employed thousands of highly skilled workers and were to be found in nearly every division of the country. New ordnance factories were built in the Midlands, North-East England, Scotland and Wales. This required workers from other parts of the country to be transferred to the new sites. The demand for new homes at these locations stimulated the local building trades. The chemical industry developed alongside the demand for armaments. A new ammonia factory was started in South Wales, and in Scotland an important new chemical works for the manufacture of ammonia by the direct combination of nitrogen and hydrogen was nearing completion. In the South, a new plant for the manufacture of sulphuric acid by means of a special vanadium catalyst was a further addition to the industry. The glass industry introduced a new method in the manufacture of mirrors for optical and scientific purposes and for television by volitisation in a high vacuum to obtain an optically perfect surface.

Concern for young people

Throughout this period of industrial development the Factory Inspectors maintained a special interest and concern for the welfare and safety of children. The permitted age of employment now protected children but young persons were vulnerable for a number of reasons, particularly when exposed to the hazards of machinery without adequate training or supervision. Mr A.R. Wilson, Chief Inspector of Factories, highlighted this concern in his Annual Report for 1936. The report commented upon the disparity between the accident rate for young workers under the age of eighteen and adults in the same occupation. Examples of accidents, given below, reinforced the opinion of inspectors about the need for special protection for young persons:

1. Boy aged 14, killed on shafting while attempting to replace a belt pulley. He had been told to do so.
2. Boy aged 14, killed by being caught by projecting shaft-end in attempting to retrieve a hammer.
3. Boy aged 15, killed by being drawn up to an overhead shaft by a belt. He had been swinging on the loop formed by a knot in the belt, which was hanging from the shaft.
4. Boy age 15, killed by falling into a pit containing a rope drive through a defective railing.
5. Two boys aged 14 and 16, found dead in hoist cage after having been trapped.
6. Boy aged 14, killed after a few days experience on a slotting machine through his shirt-sleeve becoming caught by a small circular saw which severed a vein in his armpit.
7. Boy age 16, lost part of a finger when operating a treadle-driven guillotine. The firm had been previously warned to provide a guard. A few months later in the same factory, a girl age 16 lost part of a finger when working on an unfenced power press.
8. Boy aged 14, lost a finger on a badly guarded press after 15 minutes so-called instruction.
9. Boy aged 17, lost two fingers and parts of two others on an unguarded press after 10 minutes experience.
10. Girl aged 15, died after being scalped through her hair becoming entangled on shafting below a bench.

11. Girl aged 17, lost three fingers on a power press. She had previously drawn the attention
of the foreman to the inadequacy of the guard.

Industrial estates and the movement of labour

The first Industrial Revolution not only depended upon the availability of machines
and power, it depended upon the availability of labour to operate the machines and
carry out the many other tasks associated with the needs of the new industries. At
that time much of the available labour was located in the south and this resulted in
a migration to the north where the textile industry was located. A similar logistics
problem existed as the new industries of the twentieth century developed but in this
case the location of these industries had gravitated to the south and the migration
of labour was reversed. This movement, which was essential to sustain the new
industrial development, was at its greatest between the two world wars.

The brief recovery of the economy in 1922 saw the start of the development of
new industrial areas. Two in the Reading area were partly developed by government
departments during the war and had passed into private ownership. Thirteen
factories were erected on one site, all equipped by an estate company, complete
with electricity, water, steam power, and adjacent to railway sidings. The other site
included a large paint and varnish works, a ventilation engineering works, wireless
apparatus works, a motor repair workshop and a perambulator factory. This trend
towards industrial estates continued and by 1924 there was a noticeable movement
of factories from central London to the outskirts of the city.

The development in new machinery and labour-saving appliances continued to
stimulate the building of new factories. The number of registered factories in 1924
increased to 142,494 from 139,992 and the number of workshops fell from 140,850
to 133,729. For the first time, the number of factories exceeded that of workshops
(see Table 5.1). The movement of industry from the old cramped conditions in
towns to the outskirts was highly beneficial. New factories were well laid out,
usually on a single floor and located in pleasant and healthy surroundings. The new
factories were well planned to facilitate economic production and provide for the
welfare of its workforce. They were well ventilated, with good lighting and heating,
medical services, canteens and facilities for indoor and outdoor recreation. One new
factory making artificial silk stockings was able to employ 2,000 workers. Much of
this industrial growth had located in the south of England, particularly in and around
London where the close proximity to the Port of London and good transportation
facilities to a ready market offered attractions to the industrialists.

Over a period of eight years from 1920, the number of registered factories in
the Southern Division had increased by over 3,000, leading to a scarcity of skilled
workers. One firm made arrangements not only to lodge workers for five days, but
also to pay their fares so that they could return to their homes at weekends. The
choice of sites was influenced by cheap means of transport. In the earlier days this
would have been the close proximity to a canal or railway but now it was road
transport and good communications, which meant that many new arterial roads
were attracting factory developments. This type of development was contrasted
by the severe depression in the textile trades resulting in factory closures and high

unemployment. Now it was not only the reduced demand, but also a tendency to concentrate production into fewer units and close old and uneconomic works which was contributing to the changing economic patterns of manufacture and employment in different parts of the country.

By 1930, however, the extent of the economic crisis and the depressed state of manufacture was beginning to be felt in the south. Despite the poor state of the economy, some development continued in and around London because of other favourable factors that persuaded most of the available investment to concentrate here. Industrial life in the metropolitan area was different from other parts of the country where generally there were only a few predominant trades. The greater variety of commerce and industry was an important reason why London had stood up to the depression. Many engineering firms moved to industrial estates at Edmonton, Enfield and Welwyn Garden City. Brentford on the Great West Road and North Acton also attracted new factories. Industrial development was taking place in the manufacture of light aircraft, gliders, radio equipment, metal window casements, diesel oil engines, paper board and paper bags for cement.

The glass silk industry, which originated in Austria, was established in Scotland in 1930. The material was used as an insulating medium and was spun into very fine fibres by an electrically controlled process and built into mats that were used for insulating steam pipes. The vegetable and fruit canning trade was developed, coming formerly from America and Canada. Eight new factories were built near agricultural areas in Linconshire, Cambridgeshire, Norfolk and Worcestershire.

Following the departure from the Gold Standard in 1931 and the imposition of tariffs, many foreign firms made enquiries about acquiring factories in this country, mainly in the London area. The influx of these foreign firms in 1932 meant that many articles previously imported were now made at home. 215 new foreign-owned factories that provided employment for nearly 10,000 workers were producing such things as toys, watch jewels and movements, clocks, ribbons and tiles. The recovery in the economy increased the tendency for labour to move south. The growth in employment could not be directly measured by the number of works registered because of the increasing size of the factory units. New industrial areas around London now included Park Royal, Perivale and Alperton. The Great West Road and the North Circular Road were particularly attractive to manufacturers. Further out, new works and expansion took place on the Slough Trading Estate. This trend was appreciated by the Factory Department, because it eased the administrative burden of registration and inspection. From a safety point of view the absence of hoists and overhead shafting on machines and better access and means of escape were welcome. On the other hand, the social cost of concentrating labour in these estates involved long distances to be travelled daily and a lack of a common bond and interest by the occupiers in their locality.

By 1934 the movement of industry and labour towards London had been going on for seven years. It was reported[98] that this had attracted 500,000 people, many from depressed areas. There was an increase of 27 per cent in insured workers in the South East and the South and Midlands accounted for about 50 per cent of the total insured workforce in the country. Industries showing the greatest growth were motor cars, cycles and aircraft production. The move towards London slowed down by 1935, mainly because of the difficulty in finding locations with an adequate workforce, which was compelling some employers to rely more on semi-skilled labour and apprentices.

The widespread industrial revival was now well established and companies were encouraged to look elsewhere for the location of their factories. The efforts of local and regional industrial development bodies that were formed in many parts of the country resulted in numerous moderate sized factories being established. This widened the range of economic activity in districts that hitherto had relied upon one or two staple industries. This action to induce industrialists to settle in areas of high unemployment through the establishment of trading estates resulted in the Team Valley Estate near Gateshead, which was a marshland site that was transformed by diverting the river to make way for new roads. About thirty new factories were installed on the estate. Similar sites were developed near Sunderland, Renfrew in Scotland, the Treforest Estate near Cardiff and Speke in Liverpool. By 1938 these trading estates were well established. Trafford Park in Manchester included a factory for the manufacture of asbestos sheets used for the roofing of the new style of factories. At Speke, a new area was set aside for further industrial development and at Team Valley, the occupied factories increased from twenty-nine to eighty-six during the year and the number employed there rose from 800 to 2,300. Refugees from Europe started twenty of the new enterprises in these estates. In the Hillington Estate in Renfrewshire, fifty units were in production and a further fifty-seven were under construction.

Prelude to war

The work of the Factory Department was now overshadowed by the outbreak of war in 1939. For the second time in its existence, the Department was faced with the urgent demands placed upon it by the government in support of the war effort. Exceptional statutory duties were laid upon the Inspectorate under the Civil Defence Act in relation to the provision of shelter against attack from the air. From April 1939 a proportion of the staff were occupied on this work. The Minister of Home Security in a speech to Parliament early in 1940 said that the Factory Inspectors had rendered invaluable service in collaboration with employers to ensure this work was carried out. The outbreak of the war brought a heavy influx of administration work in connection with applications to the Secretary of State for Orders under Section 150 of the Act to enable extended hours to be worked. The war perhaps inevitably saw an increase in the number of industrial accidents. The cause of the greatest increase in fatal accidents was persons falling from a height. One indication of the nation's reluctance to believe that war was inevitable was given by the number of deaths in September 1939. In spite of the warnings given by the government, little effort had been made to black out factories. Many accidents occurred in the rush to get this work done. The total of 129 deaths due to falls was the highest on record and special warnings were given to the public in the press and by radio broadcasts.

CHAPTER 6

THE FIRST ENGINEERING INSPECTORS 1921–1925

A need for specialist skills

As soon as the Inspectors of Factories were given responsibility for the safety of machinery in the mills and factories under the 1844 Act, they found themselves in conflict with the mill owners and the machine manufacturers who supplied the textile machines and the steam boilers and engines. Decisions on the practicability or otherwise of fencing a particular part of a machine against some danger to the operators required careful consideration and technical judgement on the part of the Inspectors. So long as the wording of the law was clear and prescriptive, the Inspectors' requirements had to be followed if legal sanctions were to be avoided. To some extent, this was alleviated by the ability of the owner to exercise his right of appeal against a notice and he could thereafter call upon the services of an expert of his own choice to act on his behalf in the arbitration procedures that followed. In most instances the expert would be a representative of the machine manufacturer with detailed technical knowledge of the workings of his machine and an interest in ensuring that the owner's interest in the machine and its production capacity was not damaged by adverse requirements for guarding. It is clear that the Inspectors were uneasy with this arrangement and in due course were happy to see the arbitration clauses disappearing from the legislation.

As the industries developed and the machinery grew more complex and diverse in its use it became apparent that the qualities of a good Inspector were derived from practical experience and training in some technical or manufacturing capacity. The nature of the work had changed. In addition to being vigilant law enforcement officers, Inspectors were expected to act as consultants and adviser. The Factories Acts furnished only the legal skeleton from which industrial standards of safety had to be determined, thus giving the Inspectors some liberty to flesh out this skeleton; but only if they possessed the requisite technical ability. The Annual Report of 1879 had this to say:

Employers often asked inspectors how best to fence their machinery with least hindrance to work. By his knowledge in engineering, by his observation of the working of different machines in the factory, by accumulated experience gathered in the course of his service, by understanding of Factory Acts and regulations an average inspector was able to supply employers and operators with valuable advice.

There was a limit, however, to what the average inspector could reasonably achieve in meeting this ideal concept of his ability. In due course the demanding technical difficulties in factory inspection hastened the need for specialist skills. When the Cotton Cloth Factories Act of 1889 came into force Mr E.H. Osborn, an Inspector from the Rochdale district, was given responsibility for technical content of the Act. It was no longer sufficient for an Inspector to be a law enforcement officer. The Factories Act 1895 extended the provisions of the 1889 Act to every textile factory and a new Cotton Cloth Factories Act was passed in 1897, to be replaced by the Factory and Workshop Act of 1901. In 1895, Mr H. Verney was made Junior Inspector for cotton cloth factories, and in 1899, Mr E.H. Osborn became the first Engineering Adviser. Mr W. Williams, who had been in charge of administration of Cotton Cloth Factory Acts since 1901, assumed the title of Inspector of Humid Textile Factories.

In 1903, Commander Hamilton Freer-Smith was made Superintending Inspector of Dangerous Trades and Dangerous Materials under the 1901 Act. In 1906, Mr A.P. Vaughan was appointed as Superintending Inspector for Dangerous Trades followed by Mr W.S. Smith who succeeded him in 1908. With the appointment of technically competent Inspectors from within the ranks of the Department, new specialist functions began to develop as part of the Factory Department's greater technical capability. Although the Inspectorate saw the need to develop specialist skills as part of its primary role as the law enforcement agency, it did not extend this to cover the type of detailed statutory inspections that were being built into the 1901 Act. The examination of steam boilers was a good example of the demarcation that was in place. As far back as 1858, Inspectors had recognised the valuable work being done by engineering insurance surveyors. The work done by these organisations continued alongside that of the Inspectors. This role has continued to the present day and is worthy of further consideration.

The role of the Engineering Insurance Industry

The new factories and mills being built in the cities and towns throughout the nineteenth century were only possible because of the availability of steam-driven transmission engines which provided power to the manufacturing machines. Large steam boilers were constructed to provide sufficient energy to drive the engines. There were many serious boiler explosions with sufficient destructive power to cause multiple casualties and destroy property. This caused considerable alarm and Sir William Fairbairn started a campaign to combat the danger of boiler explosions. In 1854, in response to this concern, the Manchester Steam Users Association was founded. It had the objective of reducing the number of boiler explosions by careful design and by means of independent and competent inspection by expert engineers.

This was very successful; it was claimed that within six years, out of 11,000 boilers inspected by their engineers, only eight exploded whereas 260 explosions occurred on boilers not examined. A writer in *The Engineer* in 1856 had this to say about steam power:

> We are under the domination of an ally who thinks nothing of squelching a few score of us whenever we come in its way.[33]

This statement followed an earlier report in an article of 18 April 1856, which had referred to a boiler explosion in Glasgow by which 'five hundred beings have been accidentally murdered'.[34]

Investigation of all boiler explosions became mandatory in the Boiler Explosion Act 1882. The success of the inspection of boilers by competent persons did not go without notice by the Factory Inspectors and the 1901 Act provided the opportunity to use this experience to make new provisions for use, maintenance and periodic examination of steam boilers. It is of interest to note that boiler explosions in America at the same time led to similar developments, resulting in the founding of the Steam Boiler Assurance Company in 1858. They carried out examinations with the same successful outcome. Third party inspection of steam boilers became established in law in America and continues to this day.

The effects of a shell boiler explosion in a factory

With the incorporation of statutory examination of steam boilers into the 1901 Act, the Factory Inspectorate established a common link with the work of insurance companies in the promotion of safety of plant and machinery. In the 1910 Annual Report,[80] it was noted that examination of boilers by persons of doubtful competency was still common but employers were beginning to realise the risk and inspection by one of the insurance companies was now the usual practice. An undesirable effect of the new requirement was that some occupiers appeared to be converting from steam plant to oil or gas and electrical power with the intention of avoiding their responsibilities under Section 11 of the 1901 Act. Concern still existed about the potential danger from steam boilers. The report stated that it was not uncommon to find boilers that were up to thirty years old. An example is given of one examination where the condition of the boiler was so bad that the examiner's hammer went through the boiler plate. The extreme difficulty was also pointed out of detecting certain kinds of defects, too small in size to allow an actual visual examination. Some nine years later, in 1919,[85] the Factory Inspectors reported that there were very few boilers now in use that were not insured and therefore periodically examined by expert boiler inspectors.

The Inspectors' acknowledgement of the role of the insurance industry was well illustrated in the 1930[94] report on steam boiler accidents. The report noted that surveyors of insurance companies now examined the majority of boilers. The low record of boiler explosions in the country was testament to both the efficiency of the examination and the competence of those who carried them out. The report acknowledged that all the insurance companies required a very high standard from their surveyors and insisted as a necessary qualification for this appointment on holding a First Class Board of Trade Certificate.

But the insurance industry was active in other safety-related matters. Various insurance associations promoted the increasing awareness of the need for guarding of machines. The 1919 Annual Report noted that certain insurance companies, in industries where accident insurance was done on a mutual basis, refused to insure against accidents until they were satisfied that the Factory Department requirements had been fully complied with. It was suggested in the report that insurance companies could exercise pressure on those employers who cause undue expenditure in the shape of claims owing to obsolete or inefficient safeguards on their machines. This influence exerted by the insurance industry goes back to when the Employers' Liability Act was passed in 1880. Factory Inspectors welcomed this Act when it was first enacted and continued to do so because it provided a financial sanction to support the Factories Act as an important contribution to the health and safety of workers.

The industrial scene changed rapidly during this period and with it the type and severity of the accidents. This can be seen in the accident statistics for 1920 to 1925 in Table 6.1. The 1919 report expressed concern about the number of accidents to young persons in shipyards. In the Clyde shipyards alone forty-six boys between the ages of fourteen and sixteen fell from heights and four were killed. Crane accidents were now a cause for concern. In 1920[86] it was noted that crane collapses resulted not only in injury to workmen but serious destruction of valuable plant. Investigations showed that cranes were not examined sufficiently often or carefully enough. Two years later the Annual Report[88] recorded 1,210 crane accidents including forty fatalities. By now there was a growing tendency on the part of large insurers of cranes to have them periodically examined by experts. The report noted that cranes in the hands of small employers were rarely subjected to any such skilled periodic inspection.

Year	All Accidents		One Day Reported Accidents See Notes 1 & 2.				Seven Day Accidents See Note 3.	
			Machinery in motion		Other Machinery		Non-Machinery	
	Total	Fatal	Total	Fatal	Total	Fatal	Total	Fatal
1920	138702	1404	36487	498	7199	176	95016	730
1921	92565	951	22946	288	4826	100	64793	563
1922	97986	843	24435	260	5098	104	68453	479
1923	125551	867	29968	292	6415	90	89168	485
1924	169723	956	35449	259	13320	195	120954	502
1925	159693	944	33719	233	13207	179	112797	532

TABLE 6.1: Accidents reported between 1920 and 1925

Note 1. Accidents caused by machinery in motion include:
Prime movers (other than locomotives), cranes and winches, grindstones and abrasive wheels, circular saws, lathes and power presses, locomotives and rolling stock on lines and others.
Note 2. Accidents caused by other types of machines include:
Electricity, molten metal, explosion, escape of gas or steam and others.
Note 3. Non-machinery accidents include:
Struck by falling body, person falling, struck by tool in use and others.

Wire ropes and other lifting equipment also attracted the interest of the Inspectors. By 1923 the Annual Report[89] was drawing attention to the necessity of thorough examination of all parts of a rope by a competent person. There was an increasing interest in referring such matters for the attention of the British Engineering Standards Association. A sub-committee was appointed to produce a standard for the use of wire ropes in service. Technical standards became increasingly important during this period and many were directly responsible for improvement in safety by design. The insurance industry was very influential in these matters partly because of its technical expertise and also because it already had its own standards available. For example, as far back as 1912, the National Boiler & General Insurance Company Limited of Manchester had published its own Rules for Steam Boiler Construction.

Formation of the Engineering Branch

Standards of safety had declined during the war years. Fatal accidents had increased from 1,287 in 1914 to 1,579 in 1918. But during 1919 steps were taken to achieve higher standards. A number of conferences took place with Industrial Councils and

Mr G. Stevenson Taylor

Trade Boards and a start was made in publishing pamphlets such as 'Fencing and Safety Precautions in Factories and Workshops and Protection of Hoists'. A leaflet was also issued called 'Suggestions for Rules for Safety Committees in Factories and Workshops'. Senior members of staff attended an International Labour Organization conference in Washington as advisers to British Government delegates. The need for accident prevention was identified in the Annual Report[85] which drew attention to a number of accidents that resulted from some defect in equipment or in the design of a machine such as the absence of a loose pulley, striking gear, belt hooks for overhead shafting or defectively arranged starting gear. The inability to quickly stop a machine was highlighted as a source of accident. The need for some type of emergency electrical stop to cut off power was identified as one means of improving the safety of electrically powered machinery. Mr G. Stevenson Taylor, shown above, who was to become the first Senior Engineering Inspector, wrote a technical report on the Dangerous Trades.

During 1920, continuing progress was made in reaching agreement and standardisation in safety precautions by means of industry conferences. Agreements were reached with the cotton, woollen and worsted, printing, bleaching and dyeing and tinplate trades. The Building Trade Joint Industrial Council submitted recommendations for regulations for woodworking machinery. The same council made recommendations for preventing accidents involving building operations. They suggested to the Inspectorate that the 1901 Act should be amended to remove the limitations that excluded building operations. In the 1920 Annual Report[86] mention is made of the useful work done by the Senior Engineering Inspector in visiting trade exhibitions and discussing safety problems with machine makers on the spot. As a result of one visit to the Machine Tool Makers Association, Mr Taylor was able to embody views in a memorandum to the Association. The report welcomed the co-operation between industry and the Department because this could do more to improve machine safety than legislation. Mention was made of an important inquiry into the grinding of metals carried out by an inspector, Mr Macklin. Pamphlets issued during the year included 'Ventilation of Factories and Workshops', 'The Use of Chains and other Lifting Gear', 'Fencing and Safety Precautions for Cotton Spinning and Weaving'. In addition, memoranda were issued on 'The Examination and Testing of Drying Cylinders' and 'The Dangers connected with the use of Acetylene Gas'.

During the year, another investigation into the organisation of the Department was carried out by a Factory Staff Committee, which recommended that the scientific and technical staff should be strengthened to cope with the new developments concerning health and safety. The new Engineering Branch was to deal with problems of industrial safety, ventilation and other matters relevant to the Factory Act. Similar changes were agreed for the Electrical Inspectors. The Engineering Branch was to be recruited from those general inspectors who had sufficient engineering experience. The new organisation of the Factory Department, as recommended by the Factory Staff Committee, took place on 1 August 1921. In response to demands for economy, the overall staff level of the Department was reduced from 235 to 211. Earlier that year, in January 1921, the new Engineering Branch was established with the following appointments:

Mr G.S. Taylor – Senior Engineering Inspector (appointed 1 April 1920)
Mr L.C. McNair (Class Ib) – Engineering Inspector
Mr C.W. Price (Class Ib) – Engineering Inspector
Mr E.L. Macklin (Class II) – Engineering Inspector
Mr C.F. Hunter (Class II) – Engineering Inspector.

The Annual Report for 1921[87] was quick to acknowledge the good work being done by the new branch and the newly appointed Engineering Inspectors:

Since its creation in January 1921 the Engineering Branch has proved its usefulness. Inspectors have been allocated subjects to specialise in. Technical work is divided and overlapping avoided. The Engineering Inspectors have been engaged in useful inquiries. They have drafted a code of regulations for chemical works, the manufacture of aerated water and the handling of hides and skins by Mr McNair, the manufacture of India rubber (Mr Price), electric accumulators (Mr Price and Dr Bridge), grinding of metals and cleaning/dressing castings (Mr Macklin). Mr Hunter has made special enquiry regarding the use and misuse of cranes and the dangers in the repair to oil tank steamers. He has also enquired into the competence of certain local persons in the Luton district to examine steam boilers in hat factories. As a result, warnings were given that certificates of quite a number could not be accepted owing to the inability to make certain calculations. Mr Stevenson Taylor in addition to staff responsibilities made special inquiries regarding explosions due to dust, metallic powders and volatile substances.

The Annual Report contained a special report written by Mr Price, about accidents caused by the in-running nip on rubber mixing rolls of which there had been fifty-six accidents in three years. This showed that the safeguarding methods were not effective. A fatal accident on dry cleaning was reported by Mr McNair, which showed that the deceased was cleaning a silk dress that had been taken out of the washing machine to be scrubbed with benzene on a slab and returned to the machine for rinsing when it burst into flames. A further report by Mr Hunter

Mr C.W. Price

concerned a fatal accident when the cast iron crankcase in an ammonia compressor fractured. Ammonia escaped into the room killing one and injuring two other workers. It was discovered that lubricating oil having a high freezing point was not suitable for use in the compressor. Other accidents involving the release of ammonia led to the observation that suitable respirators would have helped.

The Engineering Branch again proved itself to be of immense value in the following year. A wide range of regulations for the manufacture of India rubber, chemical works and wood-working machinery were made during the year. These regulations were made under Section 79 of the 1901 Act, and proved to be a very effective way of improving the standard of safeguarding in the respective industries. Perhaps the most important code concerned woodworking machinery, which after prolonged discussion was finally adopted. This code specified details of guards for circular saws, planing machines, band saws, spindle moulding machines and chain mortising machines all of which were intended to bring the general standards up to the level of the best equipped works. Objections were raised to all the codes, resulting in prolonged negotiations, but the objections were overcome and the codes accepted. Proposed regulations for building operations remained in draft but progress was made towards agreement following a conference with the industry. Further publication of pamphlets continued: 'Protection of Hoists (2nd Edition)', 'Use of Abrasive Wheels', 'Dangers from Acetylene Gas and the Oxy-acetylene Welding in Factories', 'Precautions in the Installation and Working of Hoists and, Precautions in the Installation and Working of Abrasive Wheels.' These publications proved exceedingly popular and were sold in large numbers.

The Annual Report of the Senior Engineering Inspector for 1922[88] contained a detailed account of the work of his Branch. Mention is made of ample flow of work in connection with the safety of various classes of machine, especially abrasive wheels, bakery machinery, cranes, centrifugal machinery, hoists, India rubber mixers, power presses, printing machines, steam boilers and other steam plant. The safe working of acetylene, compressed air, compressed ammonia and producer gas plant is also covered in the report. Various codes of regulations gave rise to special enquiries and attendance at several trade conferences. The report goes on to describe the work of the Engineering Inspectors. Mr McNair comes in for special praise for his enquiries into the safeguarding of dough mixers and dough brakes:

> With his genius for mechanical devices he has been able to suggest considerable improvements in some existing safeguards and also the development of others on entirely new lines. These have been placed on the market by manufacturers. He has made sketches of various automatic locking, feeding and stopping devices for different machines. Finished drawings are being prepared by the senior clerk, Mr E.W. Stott for the pamphlets.

Mr McNair must have been a man of considerable talent and energy. The report describes his enquiries into several explosions arising from the ignition of carbonaceous dusts such as coal, dye substances, malt and palm kernel; also from acetylene and from the use of blow lamps. He enquired into the dangers connected with dry cleaning by means of benzene and he investigated with Dr Bridge the use of chlorine gas for bleaching flour. He further devoted attention to questions arising from the new Chemical Works Regulations by visiting various works to advise occupiers on the best method to comply with the law. His expert chemical knowledge and wide experience of chemical processes was acknowledged as being invaluable.

During the year 1922 there were 97,986 accidents including 843 fatalities. Machinery accidents accounted for approximately one third of the total (29,533). The increase in the numbers over the previous year (5,421) was not seen as a fall in standards but rather as a reflection of the change in trade conditions. Nevertheless, the need for accident prevention was now an objective of the Department. This period saw the start of the policy of prevention of accidents by 'Safety First' means and recognition of the British Industrial Safety First Association. The Engineering Branch through the work of its inspectors made a great contribution to this policy. Mr Macklin, in co-operation with Dr Middleton of the Medical Branch, was responsible for the safety pamphlets on abrasive wheels. Mr Macklin was also concerned with the application of the Celluloid Regulations and visited many works around the country. He met local authorities to secure their co-operation regarding means of escape in case of fire and to discuss the application of the Cinematograph Film Act, 1922. Mr Hunter continued to devote his attention to accidents caused by the failure or collapse of cranes and failure of wire ropes. He investigated accidents caused by the failure of centrifuges, steam and gas plant, ammonia stills, refrigeration plant and the collapse of a large weaving shed in the Huddersfield district. His inquiry into the competency, or otherwise, of persons who carried out the examination of steam boilers and issued reports continued. In every case where the person could not be considered competent, they gave up the practice, and in no case was it necessary to resort to the courts for a decision.

The workload falling on the Engineering Branch in these early years was enormous and placed great demands upon the limited resources available. This is illustrated in the 1922 Annual Report[88], where Mr Stevenson Taylor records his own contribution to this work. This included the receipt of 400 papers on various subjects submitted by the Central Office staff and a similar number of papers from the District Staff asking for advice on a wide range of technical matters. In addition, the branch received all reports by the Board of Trade on boiler explosions. Mr Stevenson Taylor made many visits to factories, usually accompanied by the District Inspector or one of his colleagues. With Dr Bridge, the Medical Inspector, he visited a number of ammunition break-down works to examine the dangers to health from explosives, mustard gas, and improper handling of shells, fuses and explosive materials. The Annual Report contains his own report into the dangers in connection with repairs on oil carrying and oil fuel ships. This report followed two serious incidents; the first in 1919 on the SS *Roseleaf* at Cardiff with the loss of twelve lives, and the second in 1920 on the barge *Warwick* at Millwall with the loss of seven lives. In addition to all this, he was the Home Office representative on the wire rope committee of the British Engineering Standards Association and he was the chairman of the sub-committee drafting a standard on the use of wire ropes in service.

1923 saw a further increase of 27,565 on the previous year's accident reports including a total of 867 fatal accidents. The Senior Engineers Report[89] for the year records an increase in the workload from the district staff and a need for priorities to be set for the allocation of resources. The principal subjects dealt with during the year were identified:

> Investigation of special classes of accidents particularly with a view to securing improvements in the safe guarding and design of machinery and plant.
> The prevention of explosions due to inflammable gases, vapours and dusts, and also those due to compressed gases.
> Safety of steam boilers and other steam plant.
> Advising on schemes of general ventilation and of dust or fume removal.
> The chemical examination of the atmosphere of workplaces for injurious gases and vapours.
> The amendment of certain codes of regulations and various questions as to the application of others to different process and plant.
> Visits to industrial exhibitions.

The work of the staff was recorded in some detail. Mr McNair completed his enquiries on the dangers of using benzene in dry cleaning. With Dr Middleton, they enquired into the physical and atmospheric conditions in some parts of steel furnaces and soaking pits in which bricklayers were employed. Mr Hunter visited the makers of aerated water machines to secure improvements in the risks of accidents caused by fragments from bottles which burst during filling. Mr Stevenson Taylor investigated several incidents involving explosions. This work included reports on dangers from compressed and flammable gases, oils, and carbonaceous and others types of dust. Reference is made to a Departmental Committee on cylinders for dissolved acetylene and the first report of the Gas Cylinder Research Committee. Mention is made of explosions due to the violent chemical action between two liquids involved in chemical or allied processes. One report concerns an explosion in a large dye factory that killed two men and seriously injured four. A mixture of oleum and nitric acid used for the nitration of an organic compound was supposed to be fed into the nitration pan slowly to keep the temperature of reaction within limits. The workmen were unable to close a supply line and the plant exploded. The possibility of an explosion occurring through the ignition of a fine cloud of dust was also highlighted because of concerns that many occupiers and managers were unaware of the dangers.

The British Empire Exhibition at Wembley took place in 1924. Mr Stevenson Taylor and his staff were kept very busy carrying out inspections on all the machinery exhibits in the Palace of Industry and the Palace of Engineering. Many of the machines, which were new, had little or no provision for safeguarding the exhibitors or the general public. At the Palace of Industry the public were able to see excellent exhibits of cotton machines, working exhibits of chocolate-making machinery, tea packing, sweet wrapping, cigarette making and soap pressing – all with good fencing. In the Palace of Engineering the Inspectors found large gas and oil engines running without fencing and inadequate rails for flywheels and couplings. They found the couplings on electric motors unfenced and dangerous. Similar defects were noted in connection with flour milling, refrigeration and other machines shown in the Colonial exhibit. The report records how the majority of the exhibitors agreed to carry out modifications but others were dilatory or raised objections.

The Exhibition attracted a number of international conferences to London and the engineering staff attended as representatives of the Home Office. The 4th International Conference of Refrigeration had technical papers that greatly interested the Inspectors. The topics covered included papers on the application of liquid oxygen and liquid air and the use in refrigeration plant of hydrocarbons such as butane and propane, and toxic substances such as ethyl chloride. Much of this information was helpful to the department, which had issued a memorandum on refrigerating plant and cold storage premises. Another important conference was the First World Power Conference held at the Exhibition. Prominent engineers and scientists from the British Empire attended this event. Papers reported on developments in connection with steam generators, water and steam turbines, steam and internal combustion engines, the application of electrical power to electro-chemical and electro-metallurgical processes and illumination. Several papers on the education and training of apprentices and engineers provided an opportunity to point out the importance of educating young engineers on safe methods of working and the embodiment of safety principles in their design of machinery or plant. During the year a large number of safety pamphlets and publications were printed or re-issued:

Ship building and ship repairing accidents: Report of Committee
Prevention of accidents at docks: Report of Conference
Refrigerating plant and cold storage premises: Memorandum
Electric arc welding: Memorandum
Electricity Regulations: Explanatory Memorandum
Dry cleaning: Memorandum
Safety pamphlets: Power presses; Hoists (3rd edition); Chains and other lifting gear (2nd edition) and Abrasive wheels (2nd edition).

Accidents continued to rise with an increase of 44,172 over the previous year at 169,723 including 956 fatalities. The increase in fatal accidents coincided with those industries that had the highest level of reported accidents. Shipbuilding and construction of buildings were the highest with 103 and 104 fatalities, followed by metal conversion and rolling mills at 97, loading and unloading in docks at 88, chemicals at 57, cement, stone and clay works at 54, woodworking saw mills at 22, gas works at 21, metal founding at 28 and metal extraction and refining at 33. Deaths due to transmission shafting were up by 9 on the previous year at 58. The Senior Engineering Inspector's report[90] showed how involved were his staff in dealing with these matters. In addition to their special duties at the British Empire Exhibition and the associated conferences, the Branch dealt with a wide variety of subjects including:

Reports on the application of exemptions from certain regulations

Drafting codes of regulations

Investigation of explosions

Investigation of accidents due to the collapse of buildings and the examination of factory buildings, which appear to be dangerous

Safeguards for various classes of machinery

Investigation of dusty operations and advising on schemes of general ventilation and of dust and fume removal

Fire prevention and means of escape in case of fire

Evidence on technical matters in legal proceedings.

The individual work of the Engineering Inspectors continued to appear in the Annual Reports from the Chief Inspector of Factories. Typically, reference is made to the work of Mr McNair in the investigation of dust explosions. This is summarised in nine technical reports covering two explosions due to coal dust in cement works, two in mills due to metal getting into the disintegrator and causing sparks and others in cotton, seed, grinding and oil cake mills. He reported on four explosions in dry cleaning washing machines caused by the ignition of benzene vapour by an electrostatic charge.

1925 was an important year for the Engineering Branch and its staff. The fast pace of the work continued in response to the relentless development of many new industries. The Annual Report[91] recorded the continuing increase in the number of registered factories, now at 144,361 against a reduction in the number of registered workshops now at 128,793. Traditional rural industries such as saddlers' workshops were closing down or becoming mechanised factories. Fatal accidents were recorded at 944, out of which 539 occurred in docks and shipbuilding, construction, chemicals, metal industries and gas works. Accidents at building operations were now top of the list with 142 fatalities, followed by 93 fatal accidents in docks, 84 in shipbuilding and 70 in the metal industries. The output of safety pamphlets continued and a number were already on their third edition, indicating just how popular they were with industry. A new safety pamphlet was issued on the fencing of bake house machinery. Two technical reports, written by Mr Stevenson Taylor were concerned with serious explosions[35, 36] in petroleum and oil storage premises. Transmission machinery continued to trouble the Factory Inspectorate, there seemed to be no end to the toll of dreadful accidents caused by this type of machinery. The problem, which first attracted their attention more than eighty years previously, persisted and their frustration is well recorded in the Annual Report:

The general industrial public, as has been repeatedly pointed out in successive annual reports, are slow to recognise the fact that the harmless looking rotating shaft, once it gets a grip of a worker's clothing can speedily cause death or grave mutilation. The victim if so caught, is lucky if he escapes with only broken limbs, usually the body is held fast by the lapped clothing and the head and legs repeatedly dashed against the ceiling or adjacent wall before the shafting can be stopped. These accidents generally arise from workers approaching revolving shafting for the purpose of shipping belts, oiling bearings, or possibly for painting or whitewashing walls or ceilings in the immediate vicinity.[91]

Crane accidents totalled 3,049 with seventy-five fatalities, which was double that of any other type of machine. Repeated failures suggested serious defects in the basic design of some types of cranes. This prompted the Home Office to approach the British Engineering Standards Association with a view to standardising the design of derrick cranes. The BESA agreed to set up a committee to consider this problem. The popularity in the use of cranes also led to an increased use of wire ropes. Mr Leonard Ward reported on sixty-three cases of wire rope failure resulting in six fatal and forty-two non-fatal accidents. His analysis of the causes showed that overloading caused by shock loading and over winding of the crane contributed to one third of the accidents.

An increase in the use of power presses in the new production methods was now attracting the attention of the Inspectorate. Mr Lauder, the District Inspector for the Glasgow District, reported of a factory with hundreds of small presses in use.

The danger of approaching a rotating shaft

At his request, the firm installed American automatic guards at a cost of £5,000. Mr McNair, the Engineering Inspector, visited forty factories to give advice on guarding power presses and he attended court hearings on two occasions. He noted the increased use of automatic guards but was of the opinion that fixed guards of good design would have been better. The ideal fixed guard, he considered, was one that covered the die without obscuring the view, while preventing fingers entering the space between the die.

In November 1925, Mr Leonard Ward took up duties as Senior Engineering Inspector in the Engineering Branch. His first report acknowledged the variety of technical problems dealt with by his staff and the intense pressure they worked under. He noted that if it were not for the human interest which the problems generally entailed, the pace of their work could not be steadily maintained. He paid tribute to Mr Hunter who had died during the year. Mr Hunter had contributed to the work of the Branch on a wide range of subjects including construction, safety appliances for hoists, failure of cranes, building construction, scaffolding and the bursting of flywheels and centrifuges. The work of the other Engineering Inspectors was acknowledged as well. Mr McNair was credited with his intimate knowledge of chemical works and processes. Industry greatly respected his special skills in finding practical solutions to engineering problems and his designs of complex safety devices. Mr Macklin continued his important work on new regulations relating to the grinding industry and advised on exhaust ventilation in factories. He had carried out surveys of the manufacture of cinematography film and investigated incidents involving fire and explosions in the manufacture and use of nitro-cellulose lacquers and varnishes. Mr Price was involved in various BESA technical committees ,preparing specifications for paints and varnishes, and he was deeply involved in

Mr Leonard Ward

the problems of guarding machinery in India–rubber works and worked with the Ministry of Agriculture on the safeguarding of farming machinery.

Mr Stevenson Taylor left the Engineering Branch in 1925 to take up new duties as Superintending Inspector of the Southern Division. During the period when he was Senior Engineering Inspector from 1921, and before as an Inspector of Dangerous Trades, he had shown by his own achievements and those of his colleagues just how important was the technical aspects of their work. Their various activities as recorded in detail in the Annual Reports from 1921 to 1925 provide ample evidence of their individual and collective contribution to health and safety at a time of great industrial change. This was perhaps the finest period of achievement for the Engineering Branch and its few but deeply committed and motivated Engineering Inspectors. In thier own way their achievements mirrored the pioneering work done by the first four Factory Inspectors nearly 100 years earlier

CHAPTER 7

THE ENGINEERING BRANCH 1926–1956

The Home Office Industrial Museum

On 5 December 1927, the Home Office Industrial Museum was opened in Horseferry Road in London. This came about because of the recommendations of a Departmental Committee in 1920 and the need to convey a positive and

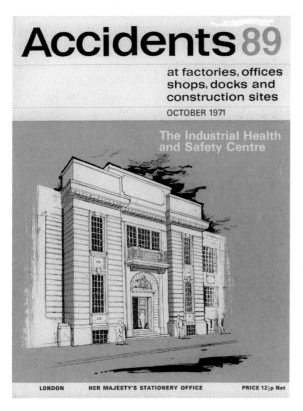

Accidents, No.89, October 1971

instructive message to industry about the diverse problems of health and safety. From its beginning the Engineering Branch was involved in designing the exhibits and in providing experienced personnel to assist in the day to day running of the museum. In the first report on the activities of the Museum, Miss Hilda Martindale, Deputy Chief Inspector, observed that members of the public who visited the museum were not mere sightseers but came with the definite purpose of adding to their knowledge. Because of this, it was imperative that experienced Inspectors were in attendance. The Museum was equally informative to visiting Inspectors from the Districts, providing valuable training on topics that they might not have experienced. In the first year, nearly 8,000 people visited the Museum, including representatives from a number of insurance companies who had called to see machines on which accidents happened. This resulted in a notice in one of the insurance journals recommending that insurance managers should visit it 'to explore its wonderfully useful collection of safety and health devices'. An illustration of the centre appeared on the front cover of 'Accidents 89' dated October 1971.

Support for the Industrial Museum was only one of the many duties undertaken by the Branch in those early years after its formation. Leonard Ward, the new head of the Branch, continued in the way of his predecessor, Mr G. Stevenson Taylor. In his report for 1928,[92] Ward describes how the efforts of the Branch had now been diverted to the improvement of the design and safeguarding of machinery to reduce the risks of accidents. He was supportive of the Safety First Movement, which was now gaining some momentum as the best course of action to reduce the high accident rates. This was reflected in the number of lectures, illustrated by lantern slides, that he and his colleagues gave on the importance of 'Safety in Industry' to technical colleges, engineering societies, insurance associations and the British Manufacturers' Association. Over 100 lectures to 16,000 persons were delivered in 1929[93] alone. The Inspectorate saw the demand for these lectures as a hopeful sign that the accident prevention message was being heeded. Reference is made to a conference at the Museum with representatives from the Association of Technical Institutions where the following conclusion was reached:

> It was most desirable particularly in large technical colleges, which are responsible for training young people about to enter all forms of industry, that they should themselves introduce into some part of the curriculum a certain amount of education in this subject.

Industrial accident statistics for this period between 1926 and 1939 are summarised in Table 7.1. In addition to the usual type of machinery accidents, other problems were coming to light with new industrial developments. Mr Pollard, the new Engineering Inspector, reported on an explosion of an oxygen cylinder, which despite being subjected to periodic examination as specified by the Gas Cylinder Research Committee, had failed because of internal corrosion. Mr McNair reported on an explosion at a chemical works in Kings Lynn where three men were killed when acetone and butyl alcohol leaked from a burst pipe. Another report involved the explosion of a pressure gauge on an oxy-acetylene welding plant and the danger of compressed oxygen coming into contact with oil and grease was highlighted. Dust explosions brought about by mechanisation and the presence of sources of ignition presented an increased hazard. Mr McNair investigated an incident in a cork mill where the dust explosion spread through the worm conveyor to an elevator and then to a cyclone dust collector, where one man received serious burns. Mr Macklin reported that the film industry had been interrupted during the previous years as the demand for 'talking pictures' has necessitated the re-organisation of plant and buildings. Another field of enquiry related to securing standards for the inspection and testing of low-pressure gasholders. This followed the wreckage of two gasholders in Manchester. Inspection of other gasholders had revealed serious deterioration in the structure, which could lead to failure. The role of the Engineering Inspectors in the investigation of these accidents is covered in more detail in Chapter 8.

Steam boilers continued to cause accidents, albeit not so frequently as during the previous decades. A serious explosion occurred at a shipyard in Birkenhead, which killed two workers. The incident was due to the weakening by corrosion of the end of the mud drums on the boiler. This had gone undetected because the external surface was concealed by brickwork. A formal inquiry by the Board of Trade found that the engineering insurance company that inspected the boiler did not think it necessary to have the brickwork removed in less than a period of twenty-five years. The inquiry reflected seriously on this but found the other inspection companies had no fixed rules on the subject. The Factory Department called a conference with all the boiler insurance companies to come up with an agreed standard for inspection. Having regard to the various conditions under which boilers were used it was recognised that some discretion must be given to the person carrying out the examination. It was agreed that it was reasonable to have the brickwork removed periodically and the following standard, which eventually was incorporated into legislation, was proposed:

> Brickwork must be removed for the purpose of thorough examination when required by the person making the examination, and in any event not less frequently than once in every six years in the case of a steam boiler situated in the open or when exposed to the weather, and not less frequently than once in every ten years in the case of a steam boiler which is properly housed and is not exposed to damp.[93]

Technical problems were not the exclusive province of the Specialist Inspectors. The 1928 report[92] refers to the work of the Superintending Inspector for Scotland, Mr Brown, and his discussions with ship owners and shipbuilders who consulted him on the requirements of the codes as they applied to shipbuilding. He observed that it was becoming common for the builders to send their draughtsmen with drawings to the Inspectorate so that points of difficulty could be discussed. Mr Stevenson Taylor in his new capacity as Superintending Inspector reported on an unusual dangerous occurrence involving the blowing over of three heavy dockside cranes at Southampton Docks during a 75mph gale. There was general concern about the safety of all types of cranes.

Crane failures in 1928 had increased by 8 per cent on the 1927 figures and a further 15 per cent increase was recorded for the following year. One way of dealing with the problem was to support the work of the British Engineering Standards Association (BSEA), in producing new specifications for the design of cranes and codes of practice for their use. The Engineering Branch allocated considerable resources to this work and Engineering Inspectors were active in thirty-eight

A sling failure on a
dockside crane

committees. Mr Pollard was elected as deputy chairman of the crane committee. There were ninety crane incidents in 1929 resulting in nineteen fatalities. The Inspectors established that insufficient examination of the cranes was an important contributory factor. They found that in eleven cases the cranes had been inspected by an insurance company, in thirty two cases the inspection was done by the crane driver or fitter but in about forty cases there had been no effective inspection or examination carried out. The Inspectors became very interested in the new development of a Safe Load Indicator, which automatically indicated the weight of the load to be lifted and gave a warning when the safe limit was reached. They were sure that the adoption of this device would do much to reduce the risk of accidents due to overloading of the crane.

An understanding of brittle fracture

The lifting and movement of heavy loads concerned the Engineering Inspectors because it was an inherently dangerous operation causing many serious industrial accidents. In 1912 a special inquiry into this problem showed the majority of

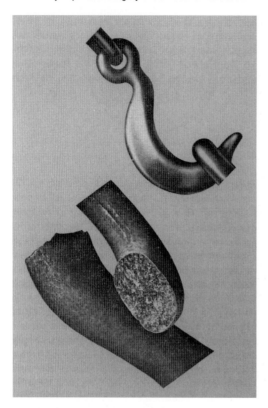

A crane hook failure

such accidents were due to a failure of chain links and the majority of fractures in the chain were due to bad welds or flaws. Legislation had to be laid down to cover this matter and from the beginning, the Inspectors were deeply involved in solving the technical challenges of making cranes and their lifting appliances safe by consideration of design and proper use and maintenance.

The only legal provision existing under the Factory and Workshop Act 1901 was in Sections 17 and 18 where the term 'plant' in Section 17 included chains. This permitted an Inspector, on finding a defective chain, to apply to a court for an order to prohibit its use altogether or until repaired. An order made under the Notice of Accidents Act 1906 required the reporting of dangerous occurrences due to the breaking of ropes, chains and other appliances operated by mechanical power in factories and workshops. The term 'other appliances' included hooks and parts of cranes, hoists, teagles and derricks operated by mechanical power whilst lifting or lowering persons or goods. Failure to report such an occurrence incurred a £10 penalty. More prescriptive requirements were introduced by the Docks Regulations 1925 and the Building Regulations 1926[38]. Regulation 19 of the Docks Regulations required testing, examination and annealing of chains and other lifting gear. Regulation 22 required means of assessing the safe working load and Regulation 23 prohibited shortening chains by knots and required the provision of packing to prevent damage to the lifting gear. Regulation 30 prohibited the overloading of the crane or its associated lifting appliances. The Building Regulations were very similar to the Docks Regulations. For any breach of these regulations, an occupier, owner or manager was liable to a penalty of £10 and any other person was liable to a £2 fine. In addition to the above two sets of regulations there existed the Shipbuilding Regulations, which were very similar to the Docks Regulations.

Mr G. Stevenson Taylor in his new position as Superintending Inspector was the author of a Home Office memorandum[41] on chains and lifting appliances published in 1929. This was a revision of an earlier one published in 1925. The memorandum dealt with the causes of accidents and the means of preventing failure or fracture of chains, rings, hooks, shackles and swivels – all essential component parts of lifting gear. It incorporated the latest research findings of the Department of Scientific and Industrial Research (DSIR) at the National Physical Laboratory. The memorandum provided information and recommendations on methods of manufacture and the type and proportion of links, rings, shackles, eyebolts, swivels, chains and slings. It emphasised the importance of the materials used in the construction of these components, which in general were wrought iron, mild steel or malleable iron. It was noted that understanding the strength of chain links involved considerable mathematical research and knowledge of the theory of strength of materials to be able to select a suitable quality. In addition, the need for inspection procedures was identified as important, particularly when dealing with small businesses who had no means of testing or sampling lifting appliances and who had to rely upon the makers trade description to determine suitability.

Of particular interest, in view of the period in which the memorandum was written, was its understanding of the effect of temperature upon the materials. Components exposed to high temperature from the heat in foundries and steelworks were considered in the memorandum. Examples were given of components attached to ladles, which were exposed to temperatures greater than 600°C, resulting in 'mysterious' fractures. One case concerned a steel ladle carried on trunnions by means of hooks, which conveyed 60 tons of molten metal to a furnace. Whilst returning empty, one hook fractured without warning, dropping the ladle. The hook was machined from special quality mild steel, which was subject to a low stress of only 2.5 tons/in^2 at the time of failure. It was concluded that it had been subject to shock at some time and this was probably aggravated by high temperature when the ladle was full. This conclusion was supported by a report in an American journal concerning similar hooks found with cracks on the sides facing the ladle, which were attributed to a combination of local heating and drilling. It was found that the heating produced a very coarse crystalline formation, resulting in the hooks becoming brittle and breaking under slight shock load when without a load.

In the case of low temperature exposure, the memorandum recorded a consensus opinion amongst chain users that chains exposed to frost became brittle and liable to fracture although at that time there was no experimental evidence to support this observation. The detection of minute cracks in metal was difficult to achieve. Current practical experience at that time showed that a fire test would facilitate this because the cracks could be seen as dark red patches when the metal was heated to a dull red. Some experienced chain smiths considered that internal defects in welds were visible by this method whilst hot water sprinkled on the components surface would help to reveal any cracks extending to the surface. This lack of a scientific understanding of this phenomenon prompted the Home Office to ask the DSIR to investigate and report their findings. Dr H.J. Gough and Mr A.J. Murray carried out test between 17°C and −78°C and provided a report in 1929[42] that concluded:

Best quality chain iron does not develop brittleness in the absence of notches within the temperature range.

Notch brittleness of chain iron commences at air temperature and increases as temperature is reduced.

Shock absorbing value of a new ¾ inch wrought iron chain of high quality material and manufacture falls slightly as temperature is lowered to −30°C. It falls at a very rapid rate as temperature drops to −78°C. This temperature brittleness is attributable to the presence of welded scarf on the link, the end of which acts as an effective notch and causes failure due to notch brittleness. Evidence shows that a link with a defect such as an imperfect weld or other surface damage may behave in a brittle manner at much higher temperatures (-10°C on test model).

Grave danger exists of brittle failure at ordinary frost temperature in chains containing links which contain defect welds or have been damaged. They should receive frequent careful inspection in order to replace defective links. Inspection or even proof testing will not reveal the separation of the weld surface due to shock load in service therefore it is desirable to allocate a definite life to a chain which is liable to be used at moderately low temperatures, depending upon service.

This memorandum covered a subject of great importance and one that continues to this day to attract the attention of those who are charged with ensuring the integrity of materials against the phenomenon of brittle fracture. It provided its readers with practical recommendations, based upon scientific research, for the prevention of failures particularly due to brittle fracture by the means of annealing and normalising periodically during the working life of the components. It also identified the need for periodic examination and testing by a competent practical man conversant with the kind of defects likely to be found in the component parts of lifting appliance. This formed the basis of more prescriptive requirements in future Factories and Workshop legislation.

The concerns about the safety of cranes and lifting appliances were well founded during this period. The Annual Report for 1931 recorded 1,498 accidents including fifty-one fatalities and an additional 1,330 dangerous occurrences. Mr L. Ward in his capacity as Senior Engineering Inspector reported on the Branch inquiries into technical questions on cranes and other lifting appliances used in building operations. As a result, the Building (Amendment) Regulations 1931 came into operation. This required major changes in the design and use of cranes. New cranes for use on building operations were to be constructed to a standard for strength and stability set out in a specification of the British Standards Institution. Existing cranes were to be brought as near as reasonably practicable to this standard by 1935 and no crane that had any timber structural member was to be used on a building operation after 31 December 1939. In the interim period, thorough examination by a competent person was to be carried out on these cranes. It was required that proper records should be kept for the testing and examination of cranes and anchorages. In addition, all cranes with a fixed or derricking jib were to be fitted with an approved Automatic Safe Load Indicator (ASLI). During that year, seven types of ASLI were in the process of being approved by the Engineering Branch. This was such an important subject that a special section was set up in the Industrial Museum to provide information on safety of lifting equipment and lifting operations.

Safe lifting equipment was vitally dependent upon the quality of the materials of construction. This was a particular problem in the early post war years and was referred to in the 1945 Annual Report[109] by Mr L.N. Duguid, who was now the Senior Engineering Inspector in charge of the Engineering Branch. His report concerned the difficulty in obtaining wrought iron chain material of a satisfactory quality and contained a copy of a letter sent by the Chief Mechanical Superintendent of a large works to one of the Superintending Inspectors in the Factory Inspectorate:

> For many years this company has placed orders for wrought iron chain, the material to conform to BS394. During the war material to this has been unobtainable, and still is. The result is that we now obtain a chain which stands up to the proof load, but the material is not up to the pre-war standard, apart from which, workmanship these days is very poor. The present position gives rise to endless worries in the use of chains to the many firms who use them, and the general opinion is that in time to come, use of such chains will be abolished.

Mr Duguid acknowledged that the difficulty of obtaining 'puddled wrought iron' and the shortage of competent chain smiths were the main reason for this unsatisfactory state of affairs. It was recorded that at one works where four ¾in wrought iron crane chains were used, they each broke after annealing and examination within two, five, seven and nine months respectively. This brought about the increased use of electrically welded mild steel chain for lifting purposes and the need to revise BS590 on Electrically Welded Mild Steel Chain to cover larger chain sizes. The report also recorded the Branch involvement in the recently published BS1290 on

Display on lifting appliances at the Industrial Health and Safety Centre

s and Sling Legs. This specification provided requirements and
to all who were called on to assess the safe working load of this
liance.

)artmental Committee – 1928

The work of a Home Office Departmental Committee[37] and its recommendations
was of great importance to the Specialist Inspectors. It was appointed to consider
what additions or changes were desirable to enable the Factory Department to
adequately discharge its existing duties and the further duties foreshadowed by
the Government Factory Bill of 1926. The committee held twenty meetings and
examined thirty-two witnesses. It was noted that the authorised strength of the
Inspectorate before the war was 223 but this was now reduced to 205 at an annual
cost of £160,000. Reference was made to the previous Departmental Committee,
which reported in 1920. That committee was instrumental in setting up the Technical
Branches and had contemplated a total strength of 237 Inspectors. Although that
target had not been reached, the number of Specialist Inspectors in post in 1928
stood at five Medical, six Engineering and five Electrical Inspector grades. Growth
in the work of the Department was due to the increase in Codes of Regulations for
dangerous and unhealthy industries made under S79 of the 1901 Act. These codes
had increased from twenty-four to thirty-seven and the extent to which they were
applied was summarised as follows:

> Factories subject to S79 had increased from 72,377 in 1914 to 185,448 in 1928.
> Woodworking Regulations in 1928 applied to 30,000 premises
> Grinding Metals (Miscellaneous Industries) Regulations applied to 10,000 premises
> Celluloid Regulations now applied to film manufacture and places where film stripping
> took place
> The Chemical Works Regulations were elaborate, requiring attention and skilled knowledge
> from the Inspectors
> Building Regulations with their safety provisions whereever power was used in construction
> were equally demanding of Inspectors

The committee recognised that the work entailed in these developments had fallen
mainly on the headquarters and technical staff but a considerable share had also
fallen on the general staff as well. This led to the conclusion that the staff levels
had to increase. This acknowledged that the new Factory Bill when it became law
would contain additional provisions concerning lighting, hoists and lifts, chains,
steam boilers, precautions in case of fire and the abolition of the distinction between
factories and workshops. The report summarised the work of the three technical
branches:

> Medical Branch. There has been a marked increase in special enquiries – dermatitis in
> sugar confectionery industry; poisoning from volatile solvents used in cellulose spraying
> and doping of aircraft; ulceration from chrome plating; the effects of asbestos dust on
> lungs; weight lifting in lamp and jute industries; effects of use of luminous paint . . . one

Inspector has conducted difficult inquiries regarding silicosis. Toxic substances are used to an increasing extent.

Electrical Branch. Present establishment (as per 1920) is one Senior Engineering Inspector at HQ and four Engineering Inspectors at London, Birmingham, Leeds and Glasgow. Primary duty is systematic inspection of works under the Electricity Regulations, now applying to over 100,000 premises also to 13,000 sub-stations. 17,000 premises require specialist inspection including 570 generation stations, railway companies and tramway authorities, over 2,700 other generating and sub-stations under the Factories Act.

Engineering Branch. Present establishment is one Senior Engineering Inspector and five Engineering Inspectors, stationed at HQ and available to visit any part of the country. Duties are to advise on engineering and mechanical questions affecting safety or health . . . fencing of machinery, ventilation in all branches of industry . . . safety of structures, cranes, lifting gear . . . explosions and chemical questions are investigated by the Branch. Engineering Inspectors play a leading part in investigations with regard to dangerous trades and processes, often in connection with proposed regulations − assist district staff in bringing new codes into operation. Careful analysis of accident reports with a view to tracing causes of accidents − checking precautionary methods adopted − give evidence at inquests and prosecutions − keep abreast of scientific knowledge and its application to industrial development.

The report by the committee went into great detail about the importance of the work done by the Engineering Branch. It acknowledged its work in the production of technical pamphlets and reports, noting that over 10,000 copies on 'The Use of Chains and other Lifting Gear' had been sold. Reference was made to the publication of penny booklets, written in simple language, for workmen to keep in their pocket. An example was 'The Treatment and Use of Chains', of which 20,000 copies were sold. Their conclusion was of vital importance to the Branch and provided clear recognition of the contribution it had made in the comparatively short period of eight years:

> We regard the work of this Branch as a vital element in the whole work of the Inspectorate . . . staff should be substantially increased by 5 to 11. A full time draughtsman should also be attached to the Branch. We suggest that some Engineering Inspectors might be stationed at other stations as an experiment. If successful, travel might be avoided . . . We think it important that the number of Engineering Inspectors should be sufficient to meet applications from the Divisions without delaying other work. It is equally important that every Division should fully utilise the Engineering Branch services.

The final recommendation of the committee was that the total staff level should be established at 283, including eight Medical, twelve Electrical and eleven Engineering Inspectors. Two Departmental Committees in the space of eight years had come to the same conclusions about the importance of the Specialist Inspectors to the Factory Department. It was to follow in later years that this recognition by these committees and the rapid way in which the Engineering Inspectors had assumed their responsibilities was to sow the seeds of discontent and lead to operational difficulties in their relationship with the general staff.

The Engineering Branch at work

As a result of these recommendations the Engineering Inspectors had increased from 5 to 10 by 1929 and Mr Ward was able to report that the increase had enabled the Branch to undertake work, previously neglected, on the design of plant and machinery from a safety point of view. It provided much needed resources for the time consuming work involved in serving on the many BESA technical committees. Gerald Bellhouse, HM Chief Inspector of Factories since 1922, was deeply involved in the development of the Inspectorate at a time of great change in industry and the new hazards that this created. He supported the 'Safety First Movement' and recognised the vital role played in this by his technical staff. He had this to say in his Report for 1930:[94]

> There is much of special interest in the work of the reports from the three Technical Branches, and perhaps the most striking feature is the increase in prevention work which has been rendered possible by the enlarged staff ... reports show how much is being done to render machinery and plant safe before it reaches the factory ... The fact that the number of electrical accidents remain so remarkably small, despite the great electrical developments, must be attributed largely to the work the Inspectors have been able to accomplish.

The first stage in the reorganisation of the Inspectorate following the recommendations of the Departmental Committee took place that year. Fifteen additional Inspectors were added to the Divisional/District staff and four Medical, two Engineering and two Electrical Inspectors were recruited to the Headquarters Technical Branches. Mr G. Scott Ram OBE, the Senior Electrical Inspector retired in 1930 after twenty-nine years' service. He was the first Electrical Inspector and worked single-handed for many years. He was responsible for framing the Electrical Regulations, which had remained fully applicable to the new developments that were taking place at that time. Mr H.W. Swann was appointed as the new Senior Electrical Inspector and in his tribute to his predecessor he acknowledged that one of his last duties before retirement was to select seven new Electrical Inspectors, all experienced electrical engineers, who would take up their appointments in 1931.

Accident levels for 1931 were beginning to decrease at long last, although this was as much to do with the developing industrial depression. The occupational hazard to health in the asbestos industry attracted the combined efforts of the Medical and Engineering Inspectors. It was accepted that the main safeguard in protecting the asbestos textile workers was the provision of improved ventilation and dust suppression. Mr Ward and Mr Price represented the Home Office in discussions with asbestos manufacturers. The consequences of this work are considered in more detail in Chapter 11. Mr Macklin who was also experienced in ventilation systems was consulted on new industrial developments in the manufacture and use of cellulose solutions, on the coating of linoleum with transparent lacquer and on the doping of aircraft wings by spraying methods rather than by hand. Mr Murray, one of the new Engineering Inspectors, was responsible for a new Air Conditioning Hut, installed in the Industrial Museum to give practical experience of the effects of humidity and other adverse ambient conditions in factories. In due course, Mr Murray took charge of the Museum.

Mr Ward reported that machine and plant manufacturers were increasingly approaching the Engineering Branch wishing to discuss safety features to be incorporated into new machines. He observed that if only the firms purchasing new machines would insist they be fenced to meet the Factories Act, better progress would be made in reducing the accident rates. During that year the Engineering Branch attended and provided expert evidence at forty-eight prosecutions and seventeen inquests and conducted 390 interviews with manufacturers who had requested expert advice on the safety of their products. Crane failures continued to cause particular concern despite the many efforts to improve the situation. Because of the problems of overturning faulty construction and the absence of testing, the Branch became involved in the drafting of the Building (Amendment) Regulations. These Regulations revised the 1926 regulations[38] by introducing the requirement to fit an approved automatic indicator of the safe working load for certain types of crane and requiring periodic testing and examination.

The Building (Amendment) Regulations came into force on 1 November 1931. This resulted in many enquiries concerning the interpretation of the regulations and the Engineering Branch were involved in many visits to buildings where cranes were in use. The requirement for Automatic Safe Load Indicators caused much interest. Applications for the approval of other types of ASLI were under consideration in

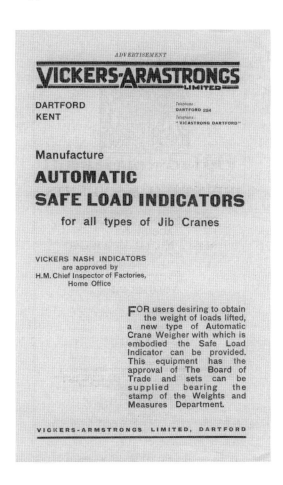

Advertisement for safety equipment

the Branch. A typical advertisement for an approved safe load indicator is illustrated on page 111. The new Shipbuilding Regulations were equally important because of their requirement for testing and examination of machinery and lifting gear. The regulations were instrumental in improving procedures for the testing and identification of hooks and chains.

1931 saw a further increase in staff levels. Fourteen Inspectors joined the Districts, the seven Electrical Inspectors selected by Mr Scott Ram took up their appointments and one Engineering Inspector joined the Headquarters staff. The accident rate continued to fall with the number of fatalities down to 755. Even the building trade recorded a reduction from 125 down to 103. Another disastrous explosion at a cordite factory accounted for twelve fatalities out of a total of forty-three in the chemical industry. Explosions in chemical works featured frequently in the Annual Reports and the special skills of Mr McNair were put to good effect. Accidents to young persons accounted for 16 per cent of all accidents and caused the Annual Report[95] to attribute this to lack of training in the operation of machines and proper instructions on their hazards. This was illustrated by the following report:

> A fourteen year old boy, two hours after he commenced to operate a drilling machine with a reaming tool in the long tapered tool holder, picked up a cotton glove belonging to a worker not using the machinery and placed it on his right hand. Almost immediately the fingers of the glove became wrapped around the revolving tool holder with the result that two fingers were literally pulled from the boy's hand.

The end of 1932 saw the completion of the first 100 years of factory legislation and the founding of the Factory Inspectorate. It was a time to review the achievements that had been made over a period of great industrial change. The Chief Inspector summarised this in his report:

> It will be seen that the last centennium has witnessed many changes in procedure, attitude and outlook in the work of the Factory Department; yet I venture to think that our present Inspectors are charged with tasks not less difficult than, though different from, those of the original nucleus of four. Ignorance, prejudice and apathy on the part of employers have indeed largely gone, yet modern industry with its ever growing complexity has tended to bring with it a stream of new and rapidly changing problems affecting the workers health and safety, each with its special risk to be examined and its special remedy to be found.[96]

The year also saw the Engineering Inspectors reach their peak of influence within the Factory Department. Both Mr Stevenson Taylor and Mr Ward were promoted to the position of Deputy Chief Inspectors in the Home Office. The importance of this is seen in the list of the Principal Officers of the Department that was published in the Annual Report for that year:

PRINCIPAL OFFICERS OF THE DEPARTMENT

Chief Inspector	D.R. Wilson CBE	Home Office
Deputy Chief Inspectors	G.S. Taylor OBE	Home Office
	Miss F.I. Taylor	
	L. Ward OBE	

Senior Medical Inspector	J.C. Bridge FRCS MRCE	Home Office
Senior Electrical Inspector	H.W. Swann AMIEE	
Senior Engineering Inspector	Vacant	

SUPERINTENDING INSPECTORS

South Eastern Division. London	H.W. Younger
Central Metropolitan Division. London	S.R. Bennett
Southern Division. London	Miss E.J. Slocock OBE
Western Division. Bristol	T.C. Taylor
Midlands Division. Birmingham	A.W. Garrett
Eastern Division. Leicester	Miss L.M.S Kelly
NortHMidlands Division. Sheffield	J. Law
North Eastern Division. Leeds	W.B. Lauder OBE
East Lancashire Division. Manchester	E.F. May
North Western Division	W. Buchan
Scotland Division. Glasgow	H.C. Thomas

The post of Senior Engineering Inspector was vacated when Mr Ward took up his new duties. It was to be more than a year before Mr Pollard took his place. Mr Ward became a Deputy Chief Inspector on the death of Mr H.J. Wilson OBE who since 1922 had been concerned with the drafting of the Building (Amendment) and Shipbuilding Regulations. The same year, 1932 also saw the death of Sir Thomas M. Legge CBE who joined the Department in 1889 and was the first Medical Inspector in 1908. After he retired in 1926, he became Medical Adviser to the TUC General Council.

Mr F.E. Pollard

Accidents continued to fall in 1932 (see Table 7.1). The continuing industrial depression was one reason for this but the Inspectorate was sure that the Safety First Movement was a factor. Building activities remained the highest at ninety-five fatalities but this was eight less than the previous year. The report was able to show that more than half of these accidents were not related to machinery. These were caused by the use of hand tools, or persons being struck by falling objects, or persons falling, or persons stepping on or being struck by objects and handling goods and articles. The human element was identified as an important factor in the cause of these accidents. Transmission machinery accidents were still too high and the Inspectorate took the opportunity of reminding the readers of the historic significance of the problem:

> Accidents are occurring at present in exactly the same fashion as they occurred nearly 100 years ago in the textile mills.[96]

The Asbestos Regulations came into force on 1 January 1932, resulting in careful attention being given to all works where this dangerous material was used in a manufacturing process. High standards of exhaust ventilation were now required and the expertise of the Engineering Inspectors was called for. Special attention was also directed to the effects of trichlorethylene, which was coming into use following the introduction of new plant for the cleaning of clothes and for degreasing engineering components. Similar use of carbon tetrachloride and the lesser use of tetrachlorethylene were also being watched.

Year	All Accidents		Machinery (See Note 1)		Machinery (See Note 2)		Non-machinery (See Note 3)		Transport (See Note 4)	
	Total	Fatal	Total	Fatal	Total	Fatal	Total	Fatal	Total	Fatal
1926	139157	806	-	-	-	-	-	-	-	-
1927	156001	973	-	-	-	-	-	-	-	-
1928	154319	953	30754	225	11078	134	110206	528	2281	66
1929	161269	962	31866	245	13601	132	113383	539	2419	66
1930	144758	899	28648	214	12246	143	101571	485	2293	57
1931	113249	755	22955	203	9438	117	78956	393	1900	42
1932	106164	602	22464	144	8928	73	73006	344	1766	41
1933	113260	688	23962	163	9767	99	77854	390	1677	36
1934	136858	780	28016	177	11787	116	94923	442	2132	50
1935	149696	843	30271	200	12126	108	105035	467	2264	68
1936	176390	920	34132	223	14794	129	124756	485	2708	73
1937	193542	1003	36405	225	15702	141	138473	568	2962	69
1938	180103	944	32769	203	13862	131	130283	528	3189	82
1939	193475	1104	34470	239	14302	136	141191	648	3512	81

TABLE 7.1. Accidents reported between 1926 and 1939.

Note 1. Accidents caused by machinery moved by mechanical power includes;
Prime movers (steam, gas and other engines, electric motors – not electric shock), transmission machinery, lifting machinery, machine tools for metal working, woodworking machinery, rollers of calenders, mixers etc. and others.

Note 2. Accidents caused by other machinery includes;
Electricity, explosions (including boiler backdraughts), fires (not dangerous occurrences or explosions), gassing, molten metal, other hot or corrosive substances, machinery not moved by mechanical power and others.

Note 3. Accidents not caused directly by machinery includes;
Use of hand tools, struck by falling body, persons falling, stepping on or striking against objects, handling goods or articles in manufacturing or carrying processes and others.

Note 4. Accidents caused by transport whether moved by power or not includes;
Railways (locomotives and rolling stock) and other vehicles (excluding hand trucks, bogies etc.).

The centenary of the Inspectorate was celebrated on 10 November 1933 by a dinner in the presence of Their Royal Highnesses the Prince of Wales and Prince George, the Duke of Kent. The guest list included Lt-Col. Rt Hon. Sir John Gilmour, the present Principal Secretary of State for the Home Office and four previous Secretaries of State. The royal guests attended because of the interest of HRH Prince George, whose official duties included gaining first-hand experience of industrial conditions by accompanied Factory Inspectors on many of their visits to factories and workshops. He continued this duty for a number of years up to the start of the war and showed a keen interest in the work of the department.

The Chief Inspector's report for that year repeated the message to its readers about the unnecessary record of death and injury that accompanied each year's account of events:

In common with predecessors, I am chiefly impressed by the industrial accidents that yearly occur, not so much by their frequency and severity, but by the number that are avoidable. The contempt for transmission machinery, which is responsible for a wholly unnecessary toll of death and disablement is almost incredible. Considering the danger has been continually emphasised from the very start of factory inspection, the present position cannot be accepted as satisfactory. Improvement remains one of the most pressing problems now before the Department. Fortunately advances in mechanical production has been associated with advances in safeguarding methods so that in many instances a machine can be so protected that an injury becomes a physical impossibility.[97]

A new method of informing the public of accident risks and the means of preventing them was introduced in the form of quarterly issues of an illustrated pamphlet called 'Industrial Accidents'. A yearly subscription to Her Majesty's Stationery Office could be obtained for 1s. These pamphlets continued for many years and provided a valuable source of information on all kinds of safety matters. The editing and

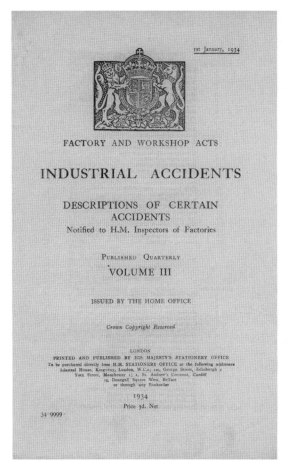

Industrial Accidents January 1934

illustration of the pamphlets was yet another duty that fell on the Engineering Branch and placed great demands upon their resources.

Other new and potentially hazardous chemicals were becoming known. One of these was dioxin (diethylene dioxide), which came to the attention of the Department following a series of deaths caused by the inhalation of vapour at an artificial silk works. The report into the investigation concluded:

The incident was followed by an inquiry into the extent this substance is in industrial use. The result was negative and the present position is reassuring in view of the public warning afforded. Subsequent research has demonstrated the toxicity of this substance but also the insidious nature of its action, and unless suitable precautions are taken there is nothing to prevent a similar catastrophe from the use of other components, the properties of which are unknown. Unfortunately there is no clear relation between toxicity and chemical constitution and having regard to the increasing use of new organic compounds as solvents, it would seem that a very valuable safeguard would be to arrange for each to be physiologically tested before it is placed on the market. It is satisfactory to report that in accordance with a decision reached more than a year ago, ICI have now taken steps for this to be done.

The importance of the new Asbestos Regulations was apparent to the Inspectorate and the long-term damage to health was only now being appreciated. This was illustrated by the following fateful assessment in the 1933 report:

> The existence of asbestosis was unrealised for many years, and the present cases of disablement through its agency are due to long past exposure to unfavourable conditions. It follows therefore that in our present state of knowledge, prevention must inevitably be preceded by the sacrifice of some health and even of life on the part of individual workers, and the main aim must be to reduce the number of such victims to a minimum.[97]

Accidents started to increase again during 1933. The general improvement in trade was given as a reason. It was also attributed to workers returning from enforced unemployment, to lack of nourishment and to being physically and mentally less alert. The metal industries took over from building work as most dangerous with ninety-one fatalities against eighty in building work. Following several fatalities in sewers, a committee was appointed by the Minister of Health to investigate and advise. Mr McNair along with Dr Middleton of the Medical Branch were appointed to represent the Department. Accidents in hoists were another area of concern for the Inspectorate. Progress in improving their safety was being retarded by so many old and out of date hoists remaining in use. This was illustrated by the following example:

> A boy fell down a hoist well from the doorway on an upper floor. The cage of the hoist was at a lower level at the time. It was not clear whether the boy found the gate at the doorway open or had opened it himself. Efficient gates on the well doorway each provided with an efficient automatic locking device so arranged that it could only be opened from the cage when at rest and that all gates must be closed and locked before the cage can be set in motion would have prevented this accident.

Transmission machinery accidents continued to frustrate the efforts of the Department and the year saw an increase in the number of fatalities from twenty-six to thirty-five. The technical solutions to many of these problems were well known but it was so difficult to get them accepted by the users. Mr W. Sydney Smith, HM Inspector of Dangerous Trades, had written a safety pamphlet[39] on the subject in 1913, as had Mr Macklin in 1925.[40] The provision of belt perches, belt mounting poles and mechanical mounting appliances all offered satisfactory solutions recommended by Inspectors. The absence of a belt perch was shown to be the cause of one such accident:

> A medium-sized power press was driven by a belt from a pulley on an overhead shaft. When in operation, the driving belt slipped off the driving pulley onto the bare revolving shaft. The belt caught on the shaft, lapped around it and was wound up, dragging the power press, which tilted and caused the belt to break. The press then toppled over on to the operator, who was incapacitated by shock and bruises.

Another example was caused by the operator being unwilling to use a belt mounting pole which if used would have kept him away from the danger zone:

A fatal accident occurred to a man who attempted to mount a belt on an overhead pulley while the shaft was in motion. It had been the practice in this factory to stop the engine for belt mounting, but apparently on this occasion the man tried to put the belt on when the shaft was running. His clothing was caught and he was carried round the shaft and he was killed.

The department was hopeful that modern improvements in machinery design would eventually reduce the risk of this type of accident. Electric motors driving individual machines, distant means of lubricating bearings and careful periodic examination of driving belts to reduce the incidence of breakage were seen as sound technical solutions to these problems.

In 1934, four major disasters in the chemical industry accounted for twenty-five fatalities, one of which killed four boys aged fifteen. The incidents were put down to the increase in industrial activity and explained away as the price that has to be paid for the increased employment and prosperity. It was noted that when the past was reviewed, the closeness between the various indices of industrial activity and accidents was remarkable. Transmission machinery accidents had increased from thirty-five to thirty-eight. Accidents caused by hoists were reported in the Annual Report[98] because the Inspectorate were convinced that many could have been avoided if hoists had been properly constructed and fenced in accordance with modern safety practices. The Engineering Branch took part in a new technical committee, which had been formed by the British Standards Institution, to prepare a standard specification for hoists and lifts. Progress was being made in crane safety and a substantial reduction in number of accidents was recorded for the year. In 1931, when the new Building Regulations came out, there were eighty accidents including nine fatalities. By 1934 this had reduced to forty-four accidents and two fatalities. Workers being struck by the revolving handles of hand-operated cranes accounted for more than half of these accidents.

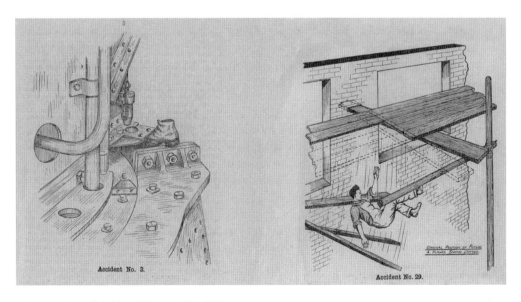

Accident No. 3.

Accident No. 29.

From *Industrial Accidents*, Volumes II and III

The success of the publication *Industrial Accidents* continued. In the two years since its existence, accounts of over 250 accidents had been published and the demand for the booklet was substantial. The publication contained simple sketches to illustrate accidents and proposed solutions to prevent their recurrence. Most of the effort for this came from the Engineering Branch. At the same time their contribution to the yearly Annual Report of the Chief Inspector ceased. In previous years the Senior Engineering Inspector wrote the chapter on safety; this year the report was written by Mr Lauder, and Divisional Superintending Inspectors wrote the chapter in subsequent annual reports. This was a starting point for the Divisions and their Districts to become more involved in the broader national issues of engineering safety. The Senior Medical Inspector and Senior Electrical Inspector continued to write their own annual account of their specialist responsibilities for many years to come. The practice of setting up conferences with industrial interests was a valuable strategy for the Factory Inspectorate. It provided a good platform on which to influence industry. Discussions took place with the Institution of Gas Engineers on the safety of pressurised gasholders, purifier boxes and the inspection and repair of water-sealed gasholders. Another conference was held with the printing, dyeing and bleaching industries to consider precautions to prevent scalding accidents at kiers, following the fatal accident involving four boys in 1934. The recommendations were accepted and agreed as the basis for a new Code of Regulations. For some years, the Department had relied on external resources for research into special problems that were outside the expertise of its own Technical Branches. The Department of Scientific and Industrial Research (DSIR) were involved in a number of projects that were recorded in the report:

Investigation of properties of iron and steel lifting tackle
Effects of electrostatic charges on inflammable substances
Detection and estimation of toxic gases and standardisation of a respirator for protection against dust
Formulation of a standing committee to advise on the precautions to be taken in connection with welded containers for compressed gases.

A special committee under the Medical Research Council was set up to investigate the toxic effects of volatile substances and the problems of inhalation of dust were being studied by a committee on industrial pulmonary diseases. The Safety of Mines Research Board were examining the causes of explosions due to industrial dust and means of prevention and the behaviour of dust particles in clouds was under examination at the University of Leeds.

During 1935, the accident rate continued to rise. The same explanation for this increase was advanced as in previous years, namely that the accident rate was related to the number of man hours worked. It was also suggested that the publicity arising from a number of important legal cases had led to a higher level of reporting of accidents to the District Inspectors. They were able to show that the perennial problem of transmission machinery accidents was reducing, albeit slowly, by comparing statistics for 1924 and 1934 as shown in Table 7.2.

Year	1924		1934	
	Total	Fatal	Total	Fatal
Shafting	312	44	164	21
Other parts of M/C's	1016	14	979	17
TOTALS	1328	58	1133	38

TABLE 7.2: Summary of transmission machinery accidents

Proceedings were taken against a textile factory in which the hoist doors were found open at three floors out of five, in spite of previous warnings. For the defence it was suggested that the human element was at fault. To which the Chairman of the Bench remarked 'If the defendant had fitted the mechanical appliances which control the human element, they would not be in court that day to plead that the human element had failed'. There were thirty-seven reported accidents to persons on or near gantries due to the movement of an overhead crane. These accidents came about because of the increasing popularity of this type of crane in factories and the nature of the accident was generally very serious or fatal as illustrated below:

> A man engaged on repairs climbed on to a girder to work near a crane track without warning the crane driver or anyone else and was fatally crushed between a roof principal and the crane.

On 3 December 1935, the Prime Minister announced that the government would introduce a new Factory Bill, for the raising and consolidation of the law relating to the safety, health and welfare of factory workers. This was welcomed by the Chief Inspector who saw it as an opportunity to deal with anomalies and defects that had been found in the 1901 Act. To some extent this was exemplified by the fact that some 377 certificates of exemption and approvals had been issued by the Chief Inspector under powers conferred on him by the many Codes of Regulations. Twenty-three new Inspectors were appointed during the year bringing the authorised staff level up to 254. Mr Leonard Ward OBE, the Deputy Chief Inspector and former Senior Engineering Inspector retired at the end of the year after thirty-five years service. The Annual Report[99] praised his contribution to the work of the Department and acknowledged his part in the setting up of the Home Office Industrial Museum and its equipment.

The main activity for the following year, 1936, was the gathering of information in connection with the Factory Bill, which was to be introduced in 1937. The year saw a substantial increase in the accident figures, which had gone up by 18 per cent with a corresponding increase of 9 per cent in fatalities. The increase in the workload was alleviated by the recruitment of more new Inspectors bringing the authorised staff level to 264. The new Bill was seen as a major event for the Factory Department because it was based on experience extending over many years. The Factory Bill was introduced as: a Bill to consolidate with amendments, the Factory and Workshop Acts 1901 to 1929 and other enactments relating to factories; and for purposes connected with the purposes aforesaid. At the same time the Education Act 1936 and the Education (Scotland) Act 1936, from 1 September 1939, raised the school leaving age from fourteen to fifteen years.

Mr Murray and Mr Price of the Engineering Branch were involved on external committees on fire brigade apparatus and heating and ventilation. Mr Stevenson Taylor retained his involvement with the ILO on the topic of prevention of accidents. Mr Taylor and another Engineering Inspector, Mr Eccles acted as advisers on safety to the building industry. Mr Eccles was subsequently to become Senior Engineering Inspector in 1947. Mr E.L. Machlin OBE, who was now promoted to the position of Superintending Inspector, wrote the chapter on safety in the Annual Report[100]. His report showed the same steady increase in accidents at 176,390 and 920 fatalities, an increase of 18 per cent and 9 per cent respectively. The increase was spread over every district though in varying degrees, the highest being London at 12 per cent, Sheffield at 25 per cent, Middlesbrough at 35 per cent, Glasgow at 36 per cent, Lanarkshire at 40 per cent and Gateshead at 50 per cent. Some factors influencing the incidence of accidents were believed to be the speeding up of processes, increased mechanisation in order to reduce production costs to meet foreign competition and the influx of new workers. In the woollen and worsted industries 15 per cent of all the accidents were caused by cleaning

Mr Eccles, Senior Engineering Inspector from 1947

machinery while it was in motion – an accident as old as industry itself – 'the old school accept industrial risks as inevitable while with boys a contempt for danger is part of their heritage'.

1937 was an important year with the introduction of the Factories Act 1937, which was passed on 30 July. The Chief Inspector introduced the new Act in his Annual Report:[100]

> The Act is a striking innovation in factory legislation. Hitherto the elasticity essential in view of the enormous variety of conditions that have to be provided for in any single Act designed to cover the whole industry has been secured by framing requirements in general terms and adapting to circumstances of each case. In the new Act an alternative method has been adopted. The requirements (especially on Safety) have been made precise and detailed. Latitude has been obtained by powers of exemption and of imposing additional restrictions where called for. This imposes both higher responsibility and greater powers to the Secretary of State and his advisers . . . Its acceptance by Parliament is a tribute to the confidence felt by industry that these responsibilities and powers will be wisely exercised.

The introduction of this new legislation and the anticipated increase in the workload of the Factory Department resulted in further increase in staff levels. Twenty-three new Inspectors were appointed during the year and a further fifty-one Inspectors were authorized, to be appointed over three years. Table 7.3 illustrates the changes in the Factory Department since 1902.

The usual increase in the annual accident rate was not quite so high in 1937 although it was still rising by 9 per cent on the previous year. In Middlesbrough for some unexplained reason the non-fatal accidents had increased by 20 per cent and fatal accidents had gone up by 40 per cent. The long-standing problem of accidents on transmission machines was illustrated by details of an accident that spanned the generations:

Staff level	1902	1938
Chief Inspector	1	1
Deputy Chief Inspectors	1	4
Superintending Inspectors	5	12
District Inspectors	42	92
Women Inspectors	8	75
Other Inspectors	79	146
Medical Inspectors	1	11
Electrical Inspectors	1	11
Engineering Inspectors	1	13
TOTAL	138	290

TABLE 7.3: Factory Department levels from 1902 to 1938

In a country mill an old man was seen by an inspector mounting a belt on to an overhead pulley while the shaft was running. When the danger of such a practice was pointed out to him he replied, 'Yes, I know, my son was killed on that shaft'. Such a terrible example had failed to convince him that shafting was dangerous, at any rate to him.

There were 432 hoist accidents with twenty-nine fatalities, compared with twenty fatalities the previous year. Most of the deaths had been due to crushing between the cage and some projection or by falling down the well. It was hoped that Section 22 of the new Act would in time see improvements in this particular type of accident. Building operations showed a further increase also with 8,223 accidents and 182 fatalities compared with 6,847 and 147 for 1936. Accidents at overhead cranes also increased by 30 per cent and hope was expressed that Section 24(7) of the new Act would be effective. Paper cutting guillotines and meat mincing machines were also causing dreadful accidents to their operators, many of whom were young persons. A serious explosion was reported in a steel foundry where eight workmen received burns and three died later (see Chapter 8). There had been several other similar accidents caused by molten metal or slag coming into contact with water. An autoclave burst with tremendous violence in a chemical works. The lid weighing 12 tons was hurled a distance of 700 feet. It caused considerable property damage and injuries to employees.

Mr E.W. Murray the Engineering Inspector was now in charge of the Home Office Industrial Museum. Mr R. Daniel was secretary to the Home Office committee on factory lighting and Mr Macklin was a member of another committee on celluloid articles. Mr Stevenson Taylor, Mr Eccles and Mr Price continued in other Home Office and ILO committees. The Inspectorate was under considerable pressure during the year, principally because of the new Act that had come into force on 1 July 1938. Conferences with users and makers was now favoured as the most effective way of solving problems of safeguarding the ever increasing number of novel and complex machinery that was being brought into use. Another system that was favoured was the setting up of Special Joint Committees to cover the whole of a particular industry. Meetings had been held with representatives of cotton spinning, cotton weaving, flour milling and paper making industries. Additional new inspectors were appointed to meet this new demand, increasing the total strength to 307. The new field covered by the Act meant that the Engineering Branch were more than usually busy in connection with the application of new technical requirements and also with the various exemptions from some of the safety requirements of the Act which the Chief Inspector was empowered to grant. Over 100 certificates of exemptions and approvals were granted during the year by the Chief Inspector. The year also saw the retirement of Mr L.C. McNair one of the original Engineering Inspectors.

In 1938, for the first time in a number of years there was a welcome fall of 6 per cent in the number of accidents and 7 per cent in the number of fatalities. There was some hope that the safety requirements of the new Act were beginning to have an effect. Reported accidents on building operations on the other hand went up by 20 per cent. There had been a substantial increase in the number of falls through roofs covered in asbestos sheeting at 99 including

17 fatalities. The reporting required by the wider application of the new Act to building operations was thought to be a possible reason for the apparent increase. Action was taken to revise the existing British Standard for this type of material.

The Annual Report for 1938[102] recognised the valuable work being done by the various committees of the British Standards Institution who were engaged in preparing and revising standard specifications for steam boilers, hoists and lifts, air receivers and fittings, wire ropes, fibre ropes, shackles and automatic fuel oil burning equipment. The Branch had co-operated in the drafting of a revision of the specification for overhead cranes requested by the Crane Makers Association to standardise methods of guarding to meet the new legal requirements:

> The work of these committees on many of which the Factory Department is represented is of far reaching importance as a means of ensuring uniform standards of safety in the construction of new plant and appliances.

The Factory Department message on safety was also being well received by industry through the publication of 'How Factory Accidents Happen', the quarterly publication which first came out in 1932. It was noted that many Trade Associations were distributing each issue free of charge to their members.

It was hoped that the provision of cage gates, mid bars, interlocking devices and landing gates for hoists in which persons ride would prevent many of the accidents to young boys and girls. In one area alone of the thirty-two accidents reported, fourteen involved young people.

One cannot read of a girl suspended by the head between the moving cage floor and the hoist shaft, of a man suspended by the leg, of feet trapped between the floor and the door lintel, of persons stepping blindly into hoist shafts and falling to the bottom without feeling that of all the safety requirements of the new Act this is one which demands most attention.

The Building Regulations of 1926 and 1931 had limitations that applied only to buildings on which mechanical power was used. These limitations were removed by the 1937 Act by including every kind of work connected with new construction, maintenance, repair work and demolition. The Regulations now needed to be revised to cover the wider application of the Act. It was acknowledged that the requirements of the 1931 Regulations, which applied to cranes was now well observed and the collapse or failure of crane structures were much fewer in number. From the end of 1939, the use of cranes partly constructed of wood was prohibited and Automatic Safe Load Indicators were now provided on nearly all types of crane.

Mr E.W. Murray reported on the Industrial Museum by observing the increased mechanisation in industry was accompanied by an increased accident risk to all workers. A special feature of the year was the museum's exhibition stand at the Empire Exhibition in Glasgow. The exhibition was open for six months and over 23,000 people visited the stand. He emphasised the need to provide proper instruction to those who come into contact with machinery by reference to a verse from Rudyard Kipling's poem 'The Secret of the Machine':

But remember please, the Law by which we live,
We are not made to comprehend a lie,
We can neither love nor pity nor forgive.
If you make a slip in handling us you die!

The developing clouds of war coming from Europe had its effects upon the Factory Department.[103] After more than 100 years under the control of the Home Office, the priorities of war resulted in the Department being temporarily transferred to the Ministry of Labour and National Service on 7 June 1940. New statutory duties were laid upon the Inspectorate as part of the preparations for war. From April 1939, a proportion of the staff was allocated to implement the Civil Defence Act, which required the provision of air raid shelters in factories. The work of the BSI continued during 1939, the various technical committees were following closely many of the new technically based legal provisions of the 1937 Act. This was of considerable assistance to the Department because the standards were setting acceptable safety guidelines for industry to follow. The Engineering Branch, in addition to existing standards work, was now closely associated with committees dealing with ship cargo lifting gear, grinding wheels, gas cylinders, welded containers, protective glass for welding operations, ventilation, hydro extractors and safety boots. Two safety pamphlets were revised during the year but it was decided to discontinue the publication of How Factory Accidents Happen for the duration of the war. It was also decided to close the museum to ordinary visitors and keep staff in attendance to a minimum.

The end of 1939 saw the retirement of Mr G. Stevenson Taylor, the first Senior Engineering Inspector to be appointed in April 1920. This was the end of an important era in the life of the Factory Department, almost as important in its own way as the beginning of the Inspectorate more than 100 years before. Stevenson Taylor and his colleagues, McNair, Price, Macklin and Hunter were pioneers who had served in the Department at a critical time when the rapid growth and complexity of engineering technology was outstripping the capability of the safety legislation to control it. They showed with their energy and dedication how to solve the many challenges that came their way. During this period many Codes of Regulations were developed under the authority of an outmoded Factories Act to address problems of safety that were unresponsive to administrative action alone. Technical based regulations were shown to be effective by ensuring that new plant and equipment was properly designed with safety as a priority. Experience had shown the value of periodic examination and testing of machinery to ensure continuing fitness for purpose and thereby reduce the risk of accidents caused by structural or mechanical failure. In its own way the Factories Act of 1937 was a fitting tribute to the work of these five remarkable Engineering Inspectors.

Power presses – a special concern

Although accidents caused by transmission machinery still occupied the attention of the Inspectors throughout this period, their concerns were directed to other types of machinery. These machines came into use because of the development of electrical power and the needs of the new industries with their requirement for high levels of production. The power press was one of a number of machines that brought its operators into close and repetitive contact with dangerous moving parts. This type of machine, like the transmission machinery before, resulted in appalling accidents requiring urgent but careful consideration of safe and practical means of safeguarding. The culmination of work on this problem spanned forty-five years, ending in 1965 with the publication of the fifth report of a Joint Standing Committee, which contained a code of requirements for power press design and tool safeguarding. This report was presented to the Rt Hon. R.J. Gunter MP, Minister for Labour. The code was intended for those who design, manufacture, sell, purchase or use power presses and their safety devices. This committee was first appointed in 1940 to examine the problems of power press safety and it published its first report in 1945. A Joint Standing Committee was appointed to review progress and make further recommendations from time to time. Four reports were drafted during this period and were now finalised and codified in this fifth report as the Power Press Safety Code.[43] The code covered a number of methods of tool safeguarding, including fixed

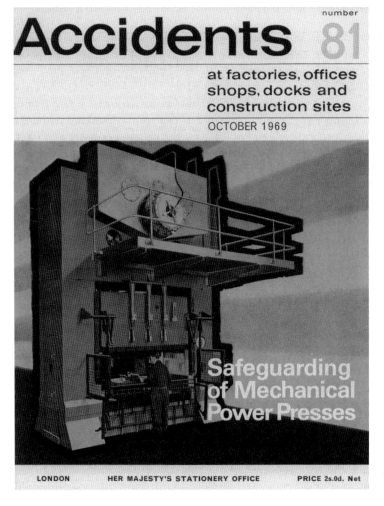

Accidents, No.81, October 1969

guards, which when in position had the merit of completely preventing access to the trapping area, always providing that their design and construction was effective. Other safety methods covered by the code included general requirements for power press arrestor brakes, electrical equipment and control for clutches, presses with friction clutches, guard construction and performance. Publication of this safety code was the culmination of many years of endeavour by the Engineering Branch in particular, to find practical solutions that would satisfy their safety objectives. Reports, going as far back as 1919,[85] clearly identified the main causes for these accidents, but the solutions were less easy to achieve. The Annual Report for that year had this to say about power presses:

> There can be little doubt that scarcely any other kind of machinery is responsible for so many accidents as power presses . . . None more perplexing than provision of safeguards, which can be suitably adapted to numerous types of presses in use and to various kinds of work. Causes of accidents are; defects in parts of the press, inefficiency of the guards and other safety devices, unsuitability of working conditions and want of care on the part of workers or supervisors.

Inspectors with technical experience of these machines were appointed in 1920 to a Joint Committee of the Tin Box Trade Board to consider safeguarding of power presses. They issued a report on new automatic guards. During that year there were 604 power press accidents with 207 occurring in the Midlands alone. It was concluded that ineffective guards gave operators a false sense of security leading to carelessness. Investigations found that several automatic guards failed to remove workers hands from the danger area because the motion of the gate was not sufficiently rapid. By 1925, Inspectors were able to report some progress in the provision of more efficient types of guards. Mr McNair from the Engineering Branch visited forty factories to give advice and attended court on two occasions to provide expert evidence for legal proceedings. The fitting of a fixed guard, whenever possible was the Inspectors' preferred solution. The ideal fixed guard was one that covered the dies while preventing fingers entering this space, without obscuring the work area.

Mr W.B. Lauder OBE, Superintending Inspector, wrote Chapter II on Safety in the Annual Report for 1933. On the question of power presses he noted that attention must again be drawn to the number of accidents that could have been prevented by more care in the selection of suitable guards or by closer supervision in their adjustment and maintenance. The third edition of the Home Office pamphlet (Safety Pamphlet No.9) was published dealing with fencing and other safety precautions for power presses. The pamphlet was completely revised to provide new and valuable advise to occupiers to assist them in the prevention of accidents. In 1939, the report of the committee examining the problems of guarding heavy power presses used in the motor industry was published. These machines had ratings in excess of 150 tons and a stroke of more than 6in at a speed of less than 30 strokes per minute. They were considered to be extremely dangerous because the operator must place the article on the dies and frequently his head was within the danger zone. With more than one operator working the machine there was the added risk of misunderstanding leading to accidents. After full consideration the committee decided that a combination of screening and mechanically operated fencing should be used. Their recommendations were illustrated in the Report of Committee (Fencing of Heavy Power Presses) published by HMSO.

This problem was further complicated by the increasing use of hydraulic presses used in the moulding of plastics. A rise in the number of accidents called for special attention on the safeguarding of this type of machine. The causes of accidents were identified as failure of hydraulic pressure in pipelines, inadvertent operation, attempts to correct an error after the machine was in motion, inadvertent placing of hands between the platens while in motion. The British Plastics Federation formed a Trade Committee to which a member of the department was co-opted.

The demands of the Second World War put great pressure upon the Inspectorate to find a balance between their long-standing concerns about the safety of these machines and the need for industry to respond to the demands of the time. The Chief Inspector's Annual Report[105] for 1941 had this to say:

> While increases of accidents are partly due to the greater use of dangerous machinery and processes in the munitions industry, they are also due to failure of the human element in their use. By this I mean a combined lack of attention on the part of the management in the provision and maintenance of guards and of the inexperienced worker in the use of these guards . . . In particular the accidents on milling machines and power presses have doubled compared with the number reported in 1938. These are machines in which new and inexperienced workers are largely employed and in addition they are machines in which the guards have to be moved or adjusted frequently and, in consequence, this work requires constant skilled supervision if it is to be done safely.

The Factory Inspectorate, throughout the war years, repeated its concerns and provided advice on the problem. In 1943,[107] it was reported that metal working machinery accidents had increased sharply since before the War. Milling machine accidents increased by 180 per cent over the 1938 figures and power press accidents by 145 per cent. This compared with an increase of 16 per cent for woodworking machines, 59 per cent for lifting machines and 73 per cent for persons falling. This concern was reflected in the Chief Inspector' Annual Report to the Rt. Hon. Ernest Bevin the Minister of Labour and National Service, which contained the following section:

> Power Presses: An analysis of the accidents in power presses brings out the depressing fact that 218 occurred last year on press tools that were not fitted with any guard whatever. This is a retrograde step compared with pre-war years; it is largely due to the fact that presses are still coming into use for production that is vital for the war effort and are put to work before a suitable guard is provided. Inspectors are waging a constant fight against this lag and we can now say that both management and the Departments interested are more helpful both as regards the labour and material necessary for the production of appropriate guards.

The report goes on to highlight the long technical fight about the advantages of fixed guards, compared with the alternatives of automatic and interlocked guards. It observed that it has become accepted practice, as contemplated by Section 14 (1) of the Factories Act, that a fixed guard must be used wherever practicable and noted that progress was being made in that direction. The necessity of safe alternatives to the fixed guard was nevertheless acknowledged and provided the justification for the Chief Inspector to set up the Standing Committee.

> Further technical progress has been made in the perfecting of interlocked guarding and I feel the initial troubles are being overcome. In fact, I now feel we have reached a stage when advances are possible that will produce a series of guards that, with proper maintenance and attention, will altogether eliminate these very serious accidents. With this faith in those technically concerned I have again asked the committees, whose work on the different aspects of the problem was suspended at the outbreak of war, to continue their work. All have agreed to do so and in addition, the Standing Committee of the heavy Power Press Committee is dealing with the question of the fencing of press (bending) brakes. I feel I must express my thanks to the members of these committees, all highly qualified in a technical sense, for the way they have agreed to my request and are now hammering out a solution to a problem that has been a source of trouble to all concerned for so many years.

The end of the war renewed the Inspectors' interest and concern about accidents on power presses. The Annual Report for 1945 recorded that these accidents were still a serious problem, which had received unremitting attention from the staff for the past twenty years. They had tackled the problem on the basis of using either fixed guards or automatic guards with a preference for the former when practicable. From a practical standpoint, it had become evident that the fixed guard, except in a relatively few cases, had not provided a satisfactory solution. It had proved difficult to obtain an adequate standard of construction, with the result that a large number of accidents had been reported each year on machines where this type of guard had been used. The automatic guard while developed to a high degree of efficiency proved unreliable for certain types of tools so that in an average press-shop the guards had been used under unsuitable conditions with the result that accidents occurred.

These considerations led to acceptance of the interlocking guard because it offered two clear advantages. Firstly, it was a type of guard the efficiency of which was not affected by the type of tool in use. Secondly, it could be arranged in conjunction with side-guards to give an effective enclosure of the danger zone. The problems associated with the interlocking guard were concerned with the measure of control it exercised over the operating parts of the press. This called for ingenuity in design and for a high standard of workmanship and discretion in deciding where it was unwise to apply this type of safeguarding. It was observed that the guard as at present designed had two objectives, i.e. to prevent clutch engagement until the guard was fully closed and to prevent opening of the guard during the motion of the crankshaft. It was acknowledged that guard makers were fully aware of these problems and were working on improvements and giving advice to users on the setting and testing of the guards. These developments were being carefully watched by the Standing Committee, which had recently been set up and had already published its first report, *Report on Safety in the Use of Power Presses*.

Power press accidents caused by trapping between the tool and die showed a substantial increase over the three-year period from 1945 to 1947. Table 7.4 illustrates the reported increase in relation to the type of guard fitted to the power

Two types of automatic guards for power presses

presses. There was particular concern in finding that one third of these accidents occurred at unguarded machines. In many cases injuries to skilled tool setters were sustained due to working when the flywheel was in motion. It was noted that in the absence of a lock on the clutch, inadvertent pedal depression would cause such incidents.

The increase in accidents related to the safeguarding of power presses led to the recommendation that a Standing Committee be set up to deal with the problems connected with the new safety devices and to advise the Chief Inspector on technical

Type of guard	1945	1946	1947
Fixed	80	122	140
Automatic	95	120	123
Interlock	35	50	91
No guard	113	157	169
Total	323	449	523

Table 7.4 Accidents relative to type of guard

developments in safeguarding these machines. There was a disappointing response to representations from the main committee to the makers of power presses to provide interlocking guards as an integral part of their machines. As a consequence, guards had to be fitted by users or the guard maker before a power press could be put into use. The Inspectors were convinced that unless the press and guard were designed as an entity a satisfactory standard of safeguarding could not be obtained. In the 1947 Annual Report note is made of the fact that power presses were being built with integral guards by a foreign manufacturer.

Table 7.5 provides a summary of all the machinery-related accidents between 1920 and 1958. It is evident that accidents caused by power presses were not a major contributor to these statistics, particularly when compared with transmission machinery and lifting equipment. One reason for this significant difference was probably due to the amount of detailed effort provided by the Inspectors and the industry to ensure that this type of machine was effectively safeguarded against the risk of serious injury to the operators. As in all cases of machinery guarding, the hazards were contained rather than being eliminated all together. This is in contrast to the very high incidence of accidents associated with lifting operations and lifting equipment. Many factors came into play here, not least being the human factor. Apart from strict statutory regulations enacted during this period, there was very little that the Inspectors could do to impose the types of engineering and supervisory control that was possible in the operational environment associated with power presses.

The central role played by the Engineering Branch in dealing with these complex machinery safeguarding problems put increasing strain upon their limited resources. In addition to the Branch's service to industry, it had to provide technical assistance to the factory inspectors out-stationed in many District Offices. This created internal difficulties which eventually had to be addressed by the Chief Inspector of Factories. In January 1955, the Minister for Labour and National Service, the Rt Hon. Sir Walter Monkton, set up a committee to examine the organisation and staffing levels of the technical branches of the Inspectorate. In the following year, a White Paper[44] was submitted to the next Secretary of State, the Rt Hon. Iain Macleod. This report examined the Factory Inspectorate and made comparisons with an earlier review[37] carried out in 1928.

The report recognised the merits of the Engineering and other Branches stationed in the Inspectorate Headquarters, but acknowledged that the needs of the non-technically skilled inspectors in the District Offices were not being met. It concluded that the Branch was heavily overworked and this work was falling into arrears. The District Inspectors were increasingly reluctant to submit work to the Branch, preferring to deal with this as best they could. It was recommended that the Branch should continue its work at Headquarters and staffing levels should be increased. The report also recommended that at least one Engineering Inspector and one Chemical Inspector should be stationed in the Divisional Offices. It took another twenty years before this recommendation was implemented. In 1978, many of the Headquarters specialist inspectors were dispersed to newly formed Field Consultancy Groups. This and other internal changes during the following years would have a fundamental impact upon the role and status of the Engineering Inspectors stationed in the Health and Safety Executive Headquarters.

Year	Prime Movers		Transmission Machinery		Power Presses		Lifting Equipment		Woodworking Machinery		All Machine Accidents	
	Total	Fatal	Total	Fatal	Total	Fatal	Total	Fatal	Total	Fatal	Total	Fatal
1920	147	5	-	-	1136	0	3271	125	2320	11	138702	1404
1921	165	5	983	42	301	3	2183	85	2424	7	92565	951
1922	214	11	938	52	384	0	2036	63	2604	6	97986	843
1923	180	6	985	49	547	1	2588	81	2258	6	125551	867
1924	202	9	1328	58	571	1	3093	76	2968	6	169723	958
1925	184	7	1363	50	609	0	3049	78	2915	7	159693	944
1928	144	4	1323	57	488	1	3388	107	3305	5	154319	953
1929	156	5	1444	47	564	0	3532	118	3300	10	161269	962
1930	132	6	1245	37	434	1	3083	104	3142	2	144758	899
1931	103	1	1006	44	379	0	2291	94	2725	5	113249	755
1932	102	7	1048	26	464	0	1927	62	2713	4	106164	602
1933	91	2	1031	18	538	0	2151	69	2754	3	113260	688
1934	112	2	1143	38	572	0	2535	86	3074	2	136853	780
1935	114	4	1257	47	613	0	2690	81	3358	3	149696	843
1936	140	2	1263	44	619	0	3267	109	3667	3	176390	920
1937	134	6	1373	39	-	-	3570	106	3594	6	193542	1003
1938	147	1	1147	32	447	0	3546	101	3554	5	180103	944
1939	167	3	1099	23	556	0	3500	123	3242	5	193475	1104
1945	87	2	861	24	608	0	4391	103	3949	6	240653	851
1946	139	3	990	24	689	0	4364	95	4266	4	223759	876
1947	197	0	935	28	665	1	4260	116	3866	5	203236	839
1948	238	5	852	19	648	1	4349	109	4020	11	201086	861
1949	230	1	792	10	529	1	4483	94	4040	7	192982	772
1950	280	0	720	21	562	0	4671	95	4162	7	192260	799
1951	328	1	656	18	548	0	4623	98	3979	7	183444	828
1952	358	2	636	9	457	0	5302	110	3649	7	177510	792
1953	299	3	635	9	418	1	4553	83	3547	7	181637	744
1954	323	2	655	14	491	1	4658	80	3475	5	185167	708
1955	353	1	596	7	502	1	4910	56	3429	4	188403	703
1956	339	4	555	9	396	0	4838	75	3229	0	184785	687
1957	264	2	548	8	426	1	4816	81	3029	2	174713	651
1958	266	0	471	7	447	0	4552	89	2926	3	167697	665

TABLE 7.5: Summary of powered machine accidents from 1920 to 1958

CHAPTER 8

FIRES, EXPLOSIONS AND DANGEROUS OCCURRENCES

Application of science to technology

Although the Industrial Revolution was made possible by the availability of the steam engine as an economic means of power, it would not have happened without the practical adaptation of scientific principles to the demands of the new engineering technologies. This new thinking in the scientific world was strongly influenced by the universities during the latter half of the eighteenth century. Edinburgh and Glasgow Universities, for example, were renowned for their teaching of science during that period. Two great chemists, William Cullen and Joseph Black, held Chairs at these universities and greatly influenced the new approach to their subject. Both men held the view that scientific study was not only an intellectual pursuit, but one that should lead to the advancement of knowledge and understanding for the benefit of society as a whole. Joseph Black, when he was Professor of Anatomy and Chemistry at Glasgow University, investigated practical matters in agriculture such as soil analysis and the effects of fertilisers.

The benefit of scientific study in the development of the steam engine is a good example of how science and technology came together at that time. The first useful industrial application of a steam engine was patented by Captain Savery in 1698 for the purpose of pumping out water from the Cornish tin mines. This design was improved by Thomas Newcomen who along with Savery obtained a joint patent for a new engine in 1705. No great improvement was made for another sixty-four years, when the weaknesses of the steam engine were deduced by James Watt. Watt was born in Greenock and worked as an instrument maker in Glasgow and London. In 1757 he was appointed to Glasgow University to repair and improve instruments in the Department of Mathematics. He became acquainted with Professor Black, who about that time published his important theory on latent heat. This led Watt to formulate his own ideas on how to improve the Newcomen steam engine. He found that heat was being wasted when the steam was injected into the cylinder under the piston for the purpose of creating a vacuum and was for that purpose condensed in the cylinder itself. The cylinder was alternately warmed by the steam and cooled by the admission of cold water to condense the steam. This cycle resulted in a great waste of steam and an equal waste of fuel to heat the steam. In 1769 he took out a patent "for lessening the consumption of steam and fuel in fire-engines" and in partnership with Matthew Boulton he developed a steam engine which condensed the steam in a separate vessel. This employed steam pressure instead of a vacuum to provide the motive force and included double-acting cylinders to give uniform motion to match the water-wheel, and most significantly, he developed mechanisms to adapt his engine to the production of rotary motion and the working of machinery.

Thus, the Age of Steam had its origins in the laboratories of Glasgow University with this early application of science to technology. Cullen and Black not only contributed to the scientific knowledge of the time, they both aimed to make science 'a study for every man of good education'. They were of a generation that was to influence the likes of Leonard Horner, who held similar views on the need to make knowledge available to the wider public. It is interesting to note that Horner and Watt shared a common interest in the literary and scientific circles of Edinburgh in the early 1800s when Watt had retired from his business. Cullen and Black encouraged people from outside the universities to attend their lectures. The advances that were to be made in industrial chemistry can be credited greatly to this open approach to their subject:

> One of a University teacher's main functions is to inspire people with curiosity; if eighteenth-century Scotland had curiosity in such good measure it was surely due in no small degree to men like Joseph Black.[45]

Allan Massie[46] observed that this curiosity in science and its practical application to the developing technologies led to the formation of the Anderson Institution at the end of the century, which was established by the will of John Anderson, Professor of Moral Philosophy at Glasgow University. Its purpose was to provide opportunities for wider education among those who were unable, for what ever reason, to follow a university course. Anderson's Institution eventually became the University of Strathclyde. Its first teacher was Thomas Garnett; he was succeeded by George Birkbeck, who took the Andersonian principle further by being responsible in 1823 for the establishment of the Mechanics' Institute in Glasgow, 'for the purpose of diffusing knowledge in literary and scientific subjects among the operatives in Glasgow.' Birkbeck moved to London where he established a similar institution in 1824 that became part of the University of London. Leonard Horner, before he became one of the first HM Inspectors of Factories, was Warden of the University of London from 1827 to 1831 and must have been a contemporary of Birkbeck with a common interest in the wider educational opportunities provided by the Mechanics' Institutions.

The start of the chemical industry

The journal *The Engineer*, first published in 1856, was quick to comment on the importance of chemistry to industry by observing 'the professional engineer has need of the deductions of chemistry'. In one of its earliest editions it quoted Professor Playfair:

It is by the researches of the practical chemist that Great Britain has raised herself to the proud position which she present holds. Baking, brewing, sugar refining, the fabrication of alkali, of glass, of pottery, of wine, of vinegar, of soap etc. are chemical operations and cannot be properly carried out without some knowledge of the principles of the science.[48]

But the chemical industry as we know it today had barely begun in 1856. Massie records[46] how Glasgow's first industrial wealth began in the early half of the century and was based upon textiles. This led to the development of ancillary industries, notably chemicals. Two industrial pioneers of that period were Charles Tennant and John White. Tennant started bleach fields in Paisley, producing a solution of chloride of lime which was an effective bleaching agent. The process was based on a French invention known as the Leblanc process, which involved the decomposition of salt by sulphuric acid. The process polluted the atmosphere with sulphurous fumes. By 1818, the Tennant family was manufacturing soda crystals and soap. He moved to Glasgow and opened the St Rollox Works, which became the largest chemical works in the world by 1825. The St Rollox Works provided bleach that reduced the cost of whitening cloth. Other industries such as glass, textiles, paper and soap benefited from the works. An attempt was made to tackle the problem of pollution by building a 435ft chimney to spread it over a wider area. Other problems associated with the Leblanc process included accumulations of solid yellow sulphurous waste, which contaminated local streams running into the River Clyde, at that time a source of drinking water:

> One paradox of Industrialisation was thus revealed: it created wealth and it contributed to a general rise in living standards – in the case of the Tennants' efforts, the production of cheap soap probably improved general standards of hygiene. However, at the same time the by-products of the industrial process were noxious.[46]

Charles Tennant's son John joined the family business as a chemist in 1815. In 1864 he established other chemical works on Tyneside. John's son Charles was the third generation to be involved in the family business and branched out on his own to set up a trading company, Tennant, Knox & Co. in London. He purchased pyrites mines in Tharsis in southern Spain. Pyrites comprised some 48 per cent sulphur. This gave him an interest in the other elements of pyrites: copper, iron, gold and silver. By 1872 he had bought seven metal extraction companies to work the ore and this led to the purchase of gold mines in India. The Cassel Gold Extraction Co. was formed and developed a process by which the amount of gold recovered from pyrites was raised, thereby improving the economics of gold mining. In association with others, John Tennant formed the Steel Company of Scotland in 1872. He also formed Nobel's Explosives Ltd to become the largest explosives company in the world where the annual output was as much as 10,000 tons in the 1880s:

> With Sir Charles's death in 1906 the Tennant connection with Glasgow weakened. Throughout the nineteenth century they had been the principal agents of wealth creation. Little of this had gone to the employees of the St Rollox Works, however. They were mostly Irish immigrants, and unskilled; the nascent trade union movement could do little with them, or for them. The health of many was ruined by the chemical processes with which

they worked. They were ill paid. In 1879 the Tennant Works paid their 3,000 employees an average of only £40 a year; Sir Charles Tennant left an estate in Britain worth more than £3 million.[46]

John and James White were the founders of another great chemical firm in Glasgow in 1810 for the manufacture of soap and soda. In 1830 it began to manufacture bichromate of potash. This was produced from chrome iron ore imported from Turkey and Russia and sold to the textile industry as an agent for fixing certain dyes such as turkey red, logwood and chrome yellow. The works employed about 500 who, like those in Tennant's, were mostly Irish immigrants. Wages were low and the working day was as long as twelve hours. According to S. Checkland:

> It was their job to do the lifting, carrying, grinding, mixing, stirring, firing the many furnaces, removing the residues, working for the most time amid great heat and noxious fumes. There was chrome furnace-men, the pearl ash-men, the crystal house- men, the workers at the vitriol tanks, and the acid towers, together with the general labourers. The chemical industry, indeed, in spite of being science based, produced the nadir of working conditions, a scene of terrible male degradation.[47]

Massie notes[46] that the Factory Inspectors found it particularly difficult to police chemical works. The employees at White's were known either as 'White's canaries' on account of the yellow dust that covered their clothes, or as 'White's dead men' owing to their faces being blanched by exposure to chemicals. The common land around the works was polluted and covered by waste dumps that leeched yellow liquid into the River Clyde. In 1899 a strike broke out at the Shawfield works and the strikers turned to the socialist leader Keir Hardie. The dispute aroused Glasgow with the revelation that the workers were paid less than 5s for a twelve-hour day. The fact that they worked in appalling conditions which led to respiratory and digestive diseases was contrasted to the comfortable lifestyle of Lord Overtoun, the chief member of the third generation of the White family, owners of the chemical works. In his defence all he could say was that his Shawfield Works had satisfied the Factory Department and the wages he paid were no lower than the going rate for unskilled labour. Massie sums up the state of the chemical industry in these early days as follows:

> The chemical industry showed industrial capitalism at its worst. Since its labour force was largely unskilled, its individual members were expendable.

In contrast to the above opinion, Disraeli was credited with saying that the chemical industry was the barometer of the nation's prosperity. Up until the 1880s Britain did lead the world in chemical manufacture but the complacency of its manufacturers meant that this position was lost to more progressive chemical industries in Germany and America. This is best illustrated in the manufacture of alkali where British manufacturers retained their interest in the Leblanc process while other countries incorporated the cheaper Solvay ammonia process. In 1874, world production of soda was 525,000 tons of which 495,000 tons were made by the Leblanc process. By 1902, production had risen to 1,800,000 tons of which the Leblanc process accounted for no more than 150,000 tons. The British Leblanc soda industry did not close down until 1920 despite the superiority of the Solvay process, which had been successfully demonstrated in this country in 1872 by Brunner Mond.

Decline and rise of the chemical industry

Almost before the British chemical industry had time to establish its position as world leader it went into decline mainly because plant owners were blind to the advantages of the Solvay process. It was reported that the development of the Solvay ammonia process had so increased the production of alkali and other chemicals that prices began to fall and the traditional chemical works on the Tyne and elsewhere were incurring financial loss. By 1883 nearly half of the alkali works on the Tyne had closed with exports to Russia, Holland and Germany in decline. The Leblanc producers were only managing to keep going on the sales of bleaching powder, which was formerly only a by-product of the process. Continuing falls in the price of these products forced the remaining companies to combine and form the Chemical Union. This failed to provide a lasting solution to the basic problem of persevering with an uncompetitive process and by 1894 large-scale unemployment was hitting the North East. Further competition to the Leblanc process took place with the development of the electrolytic process for making caustic soda and chlorine from rock salt, and the commissioning of the St Helens electrochemical works in 1896.

Germany was quick to take the lead in world chemical manufacture. The first commercial liquefaction process was developed there and superior research facilities established their position as leaders in the production of dyestuffs. The German industry success was summed up in *The Engineer* of 26 April 1907,[48] following the announcement that three companies, Bayer, BASF and Gesellschaft für Anilin Fabrikation had made between them profits of £1,500,000 in a year and declared dividends of up to 36 per cent.

> Originally these large concerns confined themselves to the manufacture of artificial dyestuffs. But they are now largely interested in pharmaceutical products, chemical foods, photographic chemicals etc. The latest development of the German chemical industry has been devoted to the manufacture of artificial manure by fixing the nitrogen of the air by an electrical method.

The First World War was the saviour of the British chemical industry. The need to be self reliant in the provision of chemicals, particularly where Germany had been dominant, meant that the industry could reform itself without the fear of foreign competition. *The Engineer* acknowledged the industry's contribution to the war effort as follows:

> Chemistry has come into its own. In war the chemist has proved himself—not supreme — but with a certain grip unknown to other men. The nations mind must be purged of the idea that chemistry has something to do with a shop, or what is called at so called public schools, 'stinks'. Does the customary person understand that this war will be won by the knowledge of the chemist.[48]

Contrary to expectations, the real growth in the chemical industry occurred in the United States. At the end of 1917 practically every chemical intermediate of importance was being produced there. The production of phenol in 1917 was more than double that of the previous year and more than 200 plants were making sulphuric acid. International competition became intense after the war and Britain's position became worse in relation to USA and Germany. In order to survive this foreign competition, four companies in 1926 agreed to merge to create Imperial Chemical Industries.

Coal, gas, oil and chemical feedstocks

While the general chemical industry was struggling because of its reliance upon the Leblanc process, the major advances in chemical technology in this country were being made in the fuel industry, mainly in gas production. Nearly all town gas in the 1850s was water gas made from Newcastle caking coal or Scottish cannel coal. Enriched coal gas with petroleum liquids was the big innovation of the 1860s because of its superior illuminating power. But large gas works located in populated areas caused fears that any fires or explosions would cause a major disaster over a wide area. The following account is given after an incident:

> During the progress of a recent conflagration many persons hastened away from the locality when they heard that there was a possibility of the flames reaching the gas-works; some thinking that if an explosion occurred even the dome of St Paul's might be endangered.[48]

In the aftermath of the war there was much debate about the economics of producing, processing and using gas, oil and coal as fuels and chemical feedstocks. The cost of importing oil products such as lubricating oil, petrol and fuel oil was compared with that of refining crude oil imports into the country. It was concluded that the possession of home-based refineries made sound economic and strategic sense. Developments took place in finding the best and most economic means of manufacturing oil products. In the 1920s, the Institution of Petroleum technologists argued that there needed to be a new British oil industry based upon the minerals found in Scotland, which would yield about 35 gallons of crude oil and 40lb of ammonium sulphate per ton by low temperature retorting. The Fuel Research Board became interested in a 'power alcohol' made by fermenting vegetable products as a substitute for petrol. Both ideas were unsuccessful.

By 1922, research had moved to the total gasification of coal to provide a cheap source of heat and power but a further goal was to produce oil from coal. The Fusion Corp. of Middlewich, in 1923, developed a retort for extracting oil from shale, torbanite, sawdust, peat and waste animal matter. A further development was the process of coal hydrogenation in which high-ash coal could be converted to oil. Strong commercial and political pressure grew to make Britain self sufficient in oil, particularly for use as fuel for the Navy and to protect the coal industry worried by cheap foreign oil. The Imperial Chemical Industries built a plant based on its own hydrogenation technology for the production of 100,000 tons per year of petrol from bituminous coal and by 1935 it was producing 4 per cent of the country's petrol needs. But the limitations were obvious and it was conceded that oil from coal could never compete with oil extracted from wells.

The renewed threat of war with Germany increased the concern for self-sufficiency in fuels. The conflicting ideas about what should be done led the government to set up the Falmouth Committee in 1937 to 'consider and examine the various processes for the production of oil from coal and other materials indigenous to this country'. Eventually the economic advantage of crude oil became too great and it became the preferred choice as a fuel and as a source of petrochemical feedstock. In 1944 the Minister for Fuel and Power appointed a committee 'to consider and report upon the effect of the hydrocarbon oil duties on the supply of raw materials to, and the development of, the chemical industry in this country'. The committee reported in 1945 and this provided the necessary stimulus to petrochemical production in Britain. They recommended that chemical manufacturers should not pay duty upon imported oil for chemical feedstock and that indigenous coal-based oil producers should be paid an amount equivalent to this duty on oils destined for chemical manufacture.

Companies were not slow to react. In 1946, Petrocarbon built a plant near Manchester for the production of olefins from naphtha or gas oil. The following year, Shell Petroleum announced their intention to produce chemicals from petroleum. It was claimed that the process would be carried out in plants containing up to date equipment, much of which had never been used in this country. In 1949 approval was given to construct an oil refinery for the Anglo-American Oil Company, later to become Esso, which would raise the national oil refining capacity by more than 500 per cent to 5 million tons per year. Shell also announced refinery expansion and in 1952, the Vacuum Oil Company, later Mobil, planned to spend £10 million on a new refinery at Coryton on the Thames Estuary. Anglo-Iranian Oil also built an installation on the Isle of Grain. From there on, the search for natural gas began onshore in 1953 and offshore in 1964, where it was discovered under the North Sea in 1965.

Legislation and the chemical industry

The chemical industry was in its infancy in 1844 when the first safety legislation for factories was being enacted. Salt, acid and alkali production for the soap and glass making industries were the main activities. Most of the developments in chemical technology were being made in the production of gas for the fuel industry. Because of this the regulation of the chemical industry was much less than applied to factories and workshops. Health and safety legislation had been fashioned over many years and was derived from a humanitarian concern for the welfare of children and young persons who were being exposed to the dangers of machinery in the manufacturing industry. The development of the chemical industry was not seen in the same light by the reformers of the day, perhaps because of the preponderance of unskilled immigrant labour and the chemical works owners' determination not to be regulated to the same degree as the factory owners. The early struggle of the chemical industry against competition from Germany and other overseas competitors also provided ample economic reason for the owners to resist the imposition of legislative control.

There is no doubt that the early Factory Act legislation did exclude chemical works from the developing health and safety law. The 1833 Act was limited to the regulation of working hours for children and young persons in mills and factories. A factory was defined as 'any cotton, woollen, worsted, hemp, flax, tow, linen or silk mill or factory wherein steam or water or any other mechanical power is used to propel or work the machinery'. The 1844 Act extended the definition of a factory to include the use of steam, water or any other mechanical power in the preparing, manufacturing or finishing or in any process incidental to the manufacture of these textiles. But the Act specifically exempted any factory used solely for bleaching, dying, printing, or calendering. The Factory Acts Extension Act 1867[49] was the first attempt to extend the definition of a factory to include premises for the manufacture India-rubber or Gutta-percha, paper, glass, tobacco, letterpress printing and book-binding. It fell to The Factory and Workshop Act, 1878[50] to set for the first time, specific safety requirements for the chemical industry to the same high standard as applied in the manufacturing industries:

> S7. Where an inspector considers that in a factory or workshop a vat, pan or other structure, which is used in the process or handicraft . . . and near to or over which children or young persons are liable to pass or to be employed is so dangerous by reason of its being filled with hot liquid or molten metal or otherwise as to be likely to be a cause of bodily injury . . . he shall serve on the occupier . . . a notice requiring him to fence the vat, pan or other structure . . . The provision of this Act to fence off machinery shall apply in a like manner as if they were re-enacted in this section with the substitution of vat, pan or other structure for machinery.

Section 31 of the same Act also required written notice to the Inspector and Certifying Surgeon of any accidents caused either by machinery moved by steam, water or other mechanical power, or through a vat, pan or other structure filled with hot liquid or molten metal or other substance. This also included the notification of any explosion or escape of gas, steam or metal and of such a nature as to prevent the injured person from returning to work within forty eight hours after the occurrence. The enforcement of this new legislation was seriously set back by the difficulty experienced by Inspectors in obtaining an appropriate penalty on conviction by the Magistrates. Typical examples of reported cases during the early 1880s were:

> 1. On 26 February 1880, R. Bealey & Co., bleachers, of Radcliffe were charged with failing to secure certain mill gearing, viz. at an upright shaft.
> They were fined £1 with £1 12s costs. The Inspector noted 'the fine appears very inadequate, a lad having been injured owing to the want of proper fencing'.[76]

> 2. On 10 August 1880, Brunner & Mond, Alkali Manufacturer, were charged with neglecting to fence mill gearing by which a person was killed on 4 June.
> The Chairman considered that there was contributory negligence by the deceased, which should be considered in mitigation. They were fined £5 with £2 8s 6d costs.[76]

> 3. On 30 March 1882, Townsend & Son, cotton and silk dyer of Coventry, were charged with omitting to report an accident which took place on the 9th and resulted in death on the 13th. They were fined 5s with 11s costs.[78]

4. On August 26 1882, The Phospho-Guano Co. of Seacombe were charged with failing to report the fatal accident to David Powell.

They were fined £5 with 16s costs. The Inspector reported – on 12 January directions were given for the large cog-wheels through which the man passed to be guarded. This had not been done when the accident occurred on 18 March.[78]

5. On 3 August 1883, The St Helens Chemical Co. of St Helens were charged with not reporting the accident of Thomas Ashall, aged fourteen through a pan containing boiling alum.

They were fined 1s with 6s 6d costs. The Inspector reported 'As there appeared to be some doubt whether defendants were not exempt under Section 100 as T. Ashall was assisting the plumber, I arranged with the defendant's solicitor to accept a nominal penalty without the case being heard.[79]

Special rules and requirements

The developing complexity of industry and the inadequacy of legal sanction were such that the Factories Act legislation still failed to address all the safety problems being found by the Factory Inspectors. To overcome this, a further revision of the Act in 1891[51] introduced in Sections 8 to 12 powers to make Special Rules and Requirements:

> S8(1) Special Rules and Requirements. Where the Secretary of State certifies that in his opinion any machinery or process or particular description of manual labour used in a factory or workshop is dangerous or injurious to health or dangerous to life or limb, or that provision of fresh air is not sufficient or that the quantity of dust generated or inhaled in any factory or workshop is dangerous or injurious to health, the Chief Inspector may serve . . . a notice in writing, either proposing such special rules or requiring the adoption of such special measures as appear to be reasonably practicable and to meet the necessities of the case.

Some ten years later in 1901[17] Parliament sanctioned the first piece of legislation that prescribed comprehensive requirements for health and safety. This included provisions for means of escape in case of fire, and powers to make orders as to dangerous machines and unhealthy or dangerous factories or workshops and powers to direct formal investigations of accidents. In Part IV of the Act covering dangerous and unhealthy industries, new requirements were laid for the reporting of occupational health matters. Medical practitioners were required to report any of their patients who they believed to be suffering from lead, phosphorus, arsenical or mercurial poisoning, or anthrax contracted in a factory or workshop. Written notice of every case of a notifiable illness occurring in a factory or workshop was required to be sent to the Inspector and Certifying Surgeon for the district. The Secretary of State also had the power by Special Order to apply these requirements to any other disease occurring in a factory or workshop.

Special Rules for chemical works were produced shortly after the 1891 Act but they were framed to meet the dangers of the Leblanc process, which was by that time being superseded by the Solvay and later on by the electrolytic processes. By 1910, the Rules were out of date and in need of revision. Two District Inspectors, Mr Jackson from Liverpool and Mr Ireland from Stockport made enquiries into the Special Rules and made the following report:

> Very few of the rules in the existing code are applicable to the ammonia-soda and the electrolytic processes, which on the other hand present dangers for which the code makes no provision (e.g., the explosion of ammonia stills etc.): the same remark applies to the manufacture of volatile hydro-carbons which has increased largely of late years. Other processes such as the manufacture of artificial indigo and aniline colours, introduced owing to the operation of the recent Patent Act, have their own special dangers. In works where the Le Blanc process is still in part carried on, the manufacturers have themselves in many cases already adopted safeguards in advance of those required by the Rules, such as the provision of rescue apparatus . . . Another explosion of chlorate of potash, similar to that at St Helens in 1899, occurred in the stores of a match factory. Since then the store house has been constructed on the plan of a "danger building" in works under the Explosives Act (1875). Smoking and use of naked lights have been prohibited; over-shoes have been provided for persons employed in the stores, and no trolleys with iron wheels are allowed inside. Experiments that have been made go to show that if even a small quantity of charred wood or carbon is allowed to fall into burning chlorate a violent explosion occurs, and for this reason it is suggested that the risk of explosion would be reduced if the material were packed in metal kegs instead of in wooden barrels.[80]

The Inspectorate supported the use of rescue apparatus in these premises. In the 1910 Annual Report it was observed that helmets fitted with an air supply or self-contained apparatus were now being provided for rescue purposes or for working in dangerous atmospheres. The Inspectors were pleased to note that periodical drill in the use by works ambulance brigade was being encouraged in some of the major gas and chemical works. The same report gave a detailed account of the special precautions that had to be taken in large gas works to prevent fires and explosions during the emptying and cleaning of gas purifiers. These purifiers were used to remove sulphurated hydrogen and other sulphur compounds from the gas before delivery to the mains. The largest purifiers contained 200 tons of ferric oxide requiring a gang of twenty-four men to empty and recharge the box in one day. The good work of the British Fire Prevention Committee was also acknowledged in the report. Reference is made to their work in obtaining data on the fire resistance of materials, systems of construction and appliances used in building practice and in the development of preventive measures to increase the protection of life and property:

> Recent reports deal with tests of fire alarm systems, fire extinguishers, fire-proof doors, and with an exhaustive series of tests made in 1909 on 'non-flam' flannelette, ordinary flannelette and 'union' flannel.[80]

Chemical works and the Engineering Branch

When the Engineering Branch was formed in 1921, it became actively involved in the safety of chemical works. One of the first tasks it undertook was the revision of the Special Rules for chemical works that were now some thirty years old and out of date. The new Branch was involved in writing the first draft code of regulations for the chemical industry. In this respect the Engineering Branch was very fortunate to have the services of Mr L.C. McNair who was one of the original four Engineering Inspectors. His contribution to safety in chemical works is covered in the previous chapters. The other Engineering Inspectors and the Senior Engineering Inspector, Mr Stevenson Taylor, also were involved in safety matters arising from other chemical related processes in connection with explosive dust and gases, ventilation, ammonia stills and explosive factories involved in the disposal of munitions.

The new Chemical Works Regulations came into force on 1 October 1922. They were technically complex in their content, ranging from the manufacture of pharmaceutical preparations to large alkali works. The Factory Department was confident that industry would be able to comply with the requirements without too much difficulty. The Inspectors did however recognise at an early stage that the Chief Inspector of Factories powers to grant exemption from certain parts of the regulations would have to be exercised. Already, exemptions in respect of the manufacture of sulphate of ammonia and dehydration of tar in gas works were under consideration.[88] Preliminary reports from the District Inspectors indicated that the majority of the large chemical works were complying with the new regulations although they were experiencing some difficulties with their interpretation. The regulations required occupiers to consider if there was a 'liability to explosion', and whether or not dangerous gas or fume was 'liable to escape'. The following year, Mr Brothers, the District Inspector for Warrington was able to report that there were ninety-nine chemical works in his district, varying from two or three persons employed, up to as many as 3,000 persons employed. In the large factories the regulations were being complied with but much work remained to be done in the smaller establishments.

The regulations contained mandatory provisions for a responsible person to examine and certify in writing that a confined space was isolated and sealed from every source of gas or fume, and was free from danger. Arising from these new requirements, experiments were being carried out by the Chemical Warfare Committee of the War Office to determine what class of self contained breathing apparatus could be sanctioned for use in gas contaminated atmospheres. There was some difficulty in granting approval for self-contained breathing apparatus although some types approved for use in mines by the Mines Department of the Board of Trade were approved for use in chemical works.[89]

The Engineering Branch made enquiries into several explosions arising from the ignition of carbonaceous dusts including coal, dye substances, malt, palm kernel and enquiries into the dangers in the use of benzene in dry cleaning. There was some concern on the part of the Department that the possibility of explosions occurring through the ignition of a cloud of fine dust of any combustible material was not sufficiently appreciated by many occupiers and managers. Explosions, caused by the chemical reaction between two liquids used in some chemical processes, were a

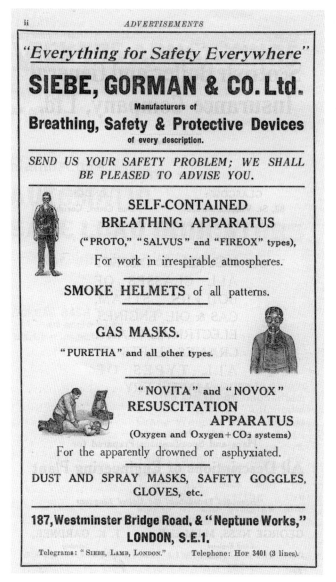
Advertisement for safety equipment

matter of concern for the Inspectorate. An explosion in 1923 caused by a mixture of oleum and nitric acid had killed two men and seriously injured four others. During this period the Inspectorate was monitoring the various technical and scientific developments taking place in the fuel industry. The 1923 Annual Report covered the work of the Joint Research Committee of the National Benzol Association and the University of Leeds on the use of highly porous bodies to recover benzol from coal or coke oven gases with particular reference to the use of silica gel.

During the following year, Mr McNair continued to assist the District Inspectors both in dealing with different chemical processes and investigating accidents and incidents involving gassing in chemical works. The Annual Report[90] reviewed the application of the regulations and provided details of fatal accidents that might have been prevented if the regulations had been observed. In one case a distillation still in which sulphureted hydrogen (H_2S) was given off was being cleaned out. Four days after cooling off the still, the foreman examined it but only applied a sense of smell test without knowing what the smell was. A man entered to start cleaning the still and collapsed. Another entered to rescue the first man and also collapsed. A third man followed with the same result. Two of the men died of gassing because they ignored the regulations, which required that before entering a confined space, they should have been equipped with suitable breathing apparatus and a life-line.

Mr Stevenson Taylor represented the Home Office at an International Congress on Refrigeration, which was held in London in 1924. New developments in this field were presented in papers on 'Liquid Oxygen Apparatus', on 'Liquid Oxygen Explosions', 'Liquid Air: Its Uses and Possibilities,' and 'Hydro-carbons and their Use in Refrigeration' covering the use of butane, propane and ethyl chloride as refrigerants. The same conference introduced an American paper, 'A Safety Code for the Refrigeration Industry' that had been drawn up by the American Engineering Standards Committee. The department noted that this code was in line with its own 'Memorandum on Refrigeration Plant and Cold Storage Premises'. It also acknowledged that this comparatively new industry recognised the importance of safeguarding its employees by the number of technical papers presented at the conference that dealt with safety.

The Factory Department continued to publish memoranda and reports on accidents happening in the chemical and related industries. In 1925 they issued a pamphlet on the storage of celluloid on the premises of professional photographers and they published two reports on accidents. One report written by Mr Stevenson Taylor and Mr Makepeace dealt with the explosion of petrol in a road tank car at the works of Shell Mex Ltd on 28 August 1924. The other, again written by Mr Stevenson Taylor dealt with the explosion of a new oil tank at the works of the Medway Oil and Storage Co. Ltd on 14 January 1925. Towards the end of 1924, an explosion occurred during the discharge of petrol from a ship in Preston Dock that killed two and injured six workers. The enquiry showed, as in previous cases, that the dangers in the handling and use of petroleum spirits were not fully appreciated and adequate precautions were not always taken to ensure the safety of workers and the protection of adjoining property. Another report dealt with an explosion of benzene vapour and subsequent fire in a raincoat proofing works, which killed three and injured five others. The process involved the passage of the fabric through a bath of benzene spirit. A spark at the commutator brush of the electric motor which drove the drying cylinder was the cause of the incident. The motor was an ordinary type with exposed commutators installed within two feet of the cylinders. The use of benzene in dry cleaning operations also caused a number of serious incidents during the year. Mr McNair provided reports on two such incidents, possibly caused by static electricity. In the first case, a silk dress had been rinsed in benzene in a washing machine. The attendant opened the machine and the vapour ignited. In the second case, a washing machine had been stopped, the material was removed and the petrol was run off when the vapour inside the wooden cylinder caught fire. On a further visit to the company, Mr McNair found three men placing a carpet, wet with petrol dripping on the floor, into a hydro-extractor.

General concern about the number of incidents of fire and explosion arising from the storage and use of petroleum spirits resulted in the enactment of the Petroleum Act in 1928, which placed new duties on the local authorities. This involved the Engineering Branch in many interviews regarding the application of the Act to factories and other premises where petroleum spirits were stored. The increasing use of lacquers in industry also attracted the attention of the branch. Mr McNair conducted tests to determine the amount of inflammable vapour present in the air in workshops and spray cabinets. This resulted in the department issuing a Memorandum on the precautions needed to reduce the risks. Mr Price, Engineering Inspector, with Mr McColgan, Electrical Inspector, also inquired into a number of fires in the spreading rooms of india–rubber works where inflammable solvents were used and where static electricity was likely to be present. Mr Pollard, Engineering Inspector, also provided a report on an unusual explosion of an oxygen cylinder, which had been subject to periodic examination and tests as specified by the Gas Cylinder Research Committee. It was found that the cylinder was badly corroded on the inside. This led to warnings about the danger when compressed oxygen comes into contact with oil or grease. A further report concerned a serious incident at a chemical works in Kings Lynn where three were killed because of leakage of acetone and butyl alcohol because of a burst pipe.[92]

The Association of British Chemical Manufacturers issued their Model Safety Rules for Use in Chemical Works in 1929. The Department welcomed the code as a valuable contribution to safety. They noted one example given in the code, in relation to the use of carbon bisulphide and a potential danger presented in the presence of pyrophoric iron sulphide. This danger had been unknown to a company using the substance, and they acknowledged that being given this data had probably prevented a serious incident occurring in their works. Dust explosions continued to attract considerable attention. During 1929, Mr McNair investigated one explosion in a cotton seed grinding plant and two other factory explosions occurred during the grinding of linseed meal and cotton seed cake. Another explosion in a cork mill spread through the worm conveyor to an elevator and to a cyclone dust collector, resulting in one man being seriously burnt by the flame. Mr McNair also enquired into explosions connected with the use of acetylene generators and a serious explosion in a sheet metal condenser connected with plant used for the sublimation of salicylic acid. It was concluded that the cloud of condensed material in the condenser was ignited either by a spark or spontaneous ignition of waste material blown into the system by hot air.[93]

During 1931, Mr Pollard reported on explosions involving compressors and air receivers. He also continued with his investigation into the safety of low–pressure gasholders, which had first caused concern in 1929. He inspected several gasholders and in one case a gasholder was found to be in such poor condition that it was condemned and dismantled. In a second case a corroded holder required a warning letter to be sent to the company demanding that it be replaced at an early date. In another incident the failure on the crown plating of a town supply gasholder occurred. A fire started, probably caused by sparks when the plating gave way. The whole of the gasholder went up in flames but no one was injured. This reinforced the Inspectors opinion on the need for regular inspection of all gasholders. Representations were

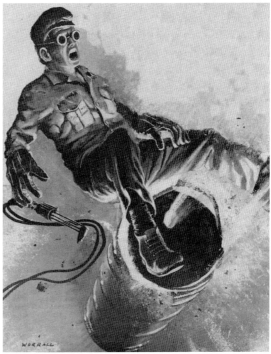

Common fire and explosion incidents
in workshops and factories

147

made to the Institution of Gas Engineers with the recommendation that periodic examinations should be carried out and systematic records maintained.

A decade of disasters

The start of the 1930s saw the country in the deepest depression when industry was operating at a low level of production; yet it proved to be a particularly disastrous period for the chemical industry, with many serious accidents. During 1930 there were a number of explosions arising from dusts, gases, vapours and other chemical compounds. One explosion at an oil cake mill belonging to J. Bibby & Sons in Liverpool killed eleven and injured thirty two workers. Another at an acid mixing plant belonging to Hickson & Partners Ltd in Castleford killed thirteen and injured fifty others. Mr McNair and Mr Peacock the District Inspector investigated the incident in Liverpool and reported:

> The explosion occurred in a large silo used for storing part boiled rice meal husk. Heating due to oxidation of seed and meal containing oil had occurred in neighbouring silos and led to spontaneous combustion in the silo in question. Steps were being taken to run out the charge from the bottom of the silo whilst applying water at the top. Two possible explanations are:
>
> 1. The collapse of some material, which had bridged on cross beams and afterwards fell onto the burning material below.
>
> 2. A small explosion of inflammable gas, due to the slow combustion of material, throwing up a cloud of dust, which ignited. [94]

The Inspectors noted that the tops of the silos were level with the working gangways on which men were employed. Above these gangways at a height of 12ft there was a concrete roof. The roof deflected flames along the gangway, resulting in the men being fatally burned. The report into the disaster included important recommendations to prevent a similar accident happening again.

Mr Newman, Engineering Inspector, and Dr Watts, HM Inspector of Explosives, investigated the disastrous explosion at Castleford. Mr Newman's report gave the following details from which it was concluded that the use of waste acid from a nitration process for making mixed acids should be regarded as a nitration process and appropriate precautions should be taken:

> The explosion occurred in a tank in which nitric acid and sulphuric acids were being mixed. Eight tons of sulphuric acid (80 per cent H_2SO_4) had been blown by compressed air from a storage tank into a mixer and nitric acid (97 per cent HNO_3) was being added from a cast iron 'egg' when evolution of nitrous fumes occurred followed by fire and explosion.
>
> It was ascertained that the contents of a railway wagon, which contained what is thought to be waste acid from one of the waste acid storage tanks, had been run as an emergency measure into the sulphuric acid tank a few days previously. Waste acid from nitration

processes always contains a small percentage of nitro-body, which gradually separates out on the surface as an oil, which should be periodically skimmed off. There is little doubt that some of this nitro-body found its way into the mixer along with sulphuric acid. The subsequent addition of nitric acid caused a violent reaction to take place resulting in the evolution of nitrous fumes and a rapid rise in temperature. [94]

This was a period of considerable activity on the part of the Factory Department in an effort to bring some level of control over the increasing number of serious incidents involving the use of chemicals. All the major investigations carried out into these incidents were issued as technical reports by the Engineering Branch and pamphlets and memoranda continued to be published. In its 1931 Annual Report the department noted that despite the application of scientific knowledge to the manufacturing processes and human care and ingenuity, explosions continued to occur, not only from scheduled explosives but also from a wide variety of solids, liquids, vapours and gases used in industry. The report also warned that petrol and other inflammable liquids were primary agents in a number of explosions; the majority occurring during soldering, welding or repairing empty tanks that had previously contained these liquids. Garages were now seen as being a particular risk and occupiers were warned of the need to free tanks of residual petrol and vapour. To add to the Inspectorate's concerns there occurred a most disastrous explosion at a nitro-glycerine department in an Admiralty cordite factory, which killed twelve and injured many others. McNair and the District Inspector reported on an explosion that killed three laboratory chemists in Stoke on Trent on a plant where spirit obtained from low temperature carbonisation of coal was being treated with sulphuric acid:

> During treatment of a batch, liquid frothed over the edge of a washer and spread along the yard where it ignited. Storage tanks were set on fire and exploded. Too concentrated acid resulted in a violent reaction with olefins present in the spirit. The risk was increased by the lack of vigorous stirring. [95]

The Thermit process for making metallic alloys caused an explosion when a warm mixture of aluminium powder and copper oxide was being mixed by shovel. This accident killed one person and seriously injured another. Mr McNair's report[95] was of considerable technical interest.

A mixture was being prepared to make copper aluminium alloy. Copper oxide was still hot and was mixed with aluminium by shovelling on an iron plate when it exploded. Experiments at the Research Department, Woolwich, found that fine dust removed from both ingredients by sifting when mixed in proper proportions could be ignited by a blow.

A number of important recommendations were made in the report:

1. The use of iron or steel implements should be prohibited.
2. Brass, aluminium or wood should be used instead.
3. Materials should be allowed to cool before mixing.
4. Fine material from aluminium should be sieved out before mixing.
5. The platform on which the materials are mixed should be dry and clean.

The following years saw the continuation of reports[96, 97] on dust explosions and other chemical related incidents. Explosions occurred in two plants in which oil cake and various seeds were ground. In the first incident four men were burnt and in the second two men were burnt. Little damage happened to the plant in both cases because explosion relief panels had been provided. Two explosions occurred in the cork grinding plant at a linoleum factory when the dust cloud in a fan duct ignited. Similar explosions were experienced on plant used for grinding wood. Following distillation of light oil in a 4,000 gallon still, an explosion bulged the bottom of the tank and damaged the brickwork. At the time of the explosion and for the preceding six hours everything in the still was closed except a small air tap. A probable explanation of the explosion was that a coating of iron sulphide on the inside of the still became incandescent owing to the oxidising action of the air and the ignition of the air-vapour mixture inside the tank. It was noted that painting the inside surface with a cement slurry to prevent corrosion would have prevented the formation of iron sulphide. A similar explosion caused by iron sulphide forming on a vulcanising plant resulted in structural damage. 1933 saw reports of coal dust explosions at a pulverising fuel plant in an electric power station and a dust explosion at a plant for grinding cattle food. In another mill that was stationary at the time, a workman was cutting out an opening in the casing of an elevator. He knocked out the plate, which disturbed dust that was then ignited by the cutting flame.

Reports on dust explosions and fire and explosions in chemical works continued. Reference was made to the explosion of coal dust at a pulverised fuel plant in a power station, which injured six employees. This was the result of an inadequate safety vent on the hopper that failed to relieve the explosion. Mr McNair with Dr Coste, Chief Chemist of London County Council, prepared a report[52] on an explosion at a synthetic chemical works, which concluded:

> A serious explosion and fire occurred at a works for the preparation of synthetic chemicals. Extensive damage was caused to the plant and adjacent houses. One boy was killed and several persons suffered shock and minor injuries. It was due to the escape of boiling methylated spirit, caused by the breakage of a bolt securing the blank flange at the back of a still. Vapour passed through the building, over a low wall towards a row of cottages behind the works where it was ignited by a domestic cooker.

Further reports are given of an explosion at a tyre works as two men were inflating an inner tube in a cast iron mould. The explosion fractured the mould and fatally injured the men. Another explosion occurred in a tar still, and several explosions were recorded when metal drums or other vessels were being repaired or cut for scrap metal. A fire in a cellulose spraying room in an aeroplane works was due to a labourer scraping deposits from a cabinet with a metal scraper. A spark caused by the scraper ignited the deposit and the fire rapidly spread to other booths.

The incidence of fires, explosions and dangerous occurrences reached a peak by 1934[98] when there were twenty-five fatal accidents in chemical works. Eleven of these occurred in an open air refrigeration plant at a chemical works where ammonia gas escaped when an aluminium bursting disc failed at a pressure which

An explosion in
a gas-drying unit
caused by an unsafe
system of work

was below the design limit of the plant. Those caught up in the incident were painters, electricians, fitters and labourers working on or around a condenser tower. The casualties suffered skin burns, intense inflammation of mucous membrane of eyes and respiratory tract. It was reported that death occurred in a few hours due to shock or acute bronchitis. The men had no reason to be aware of danger and respirators were not worn or easily accessible. In another incident four young boys aged fifteen were killed in scalding accidents at kiers in a bleach works:

> The accident occurred in Bleach Works and resulted in the death of four boys who were severely scalded by boiling liquor when plaiting down cloth in a pressure kier. The accident was due to the inadvertent admission of boiling liquor from the heater of an adjoining kier, which was being blown down. The actual presence of more than two boys in the kier was a most unusual event and was due to the addition of two boy learners to the regular plaiters.[98]

This accident led to a conference with the Printing, Dyeing and Bleaching Industries in Manchester where recommendations were made. They were accepted and subsequently incorporated into The Kiers Regulations 1938. Another conference took place that year with the Coal Burning Appliance Makers Association and Users of Pulverised Fuel Plant with the object of producing a code to prevent coal dust explosions. Despite this, the following year saw a considerable number of explosions associated with the use of coal dust.

The rupture of a crown plate on a gasholder at Barrow in Furness resulted in the ignition of a large volume of gas, which escaped as the lifts collapsed. Fortunately there were no serious injuries or damage to adjacent property. Examination of the crown plate revealed very deep internal corrosion, which weakened the plate. The explosion led to an inquiry into the effectiveness of the recommendations made by the Institution of Gas Engineers in 1932. Consideration was also given to the possibility of determining the internal state of the plates by new methods that would not require an entry into the vessel. A serious fire in a chemical works also attracted the following report:

> The accidental displacement of a screw plug at an oil cracking plant resulted in the escape of a large quantity of inflammable liquid under pressure. The vapour ignited, probably by a furnace, 20 yards away at the other side of a partition. Two men whose clothing had been saturated by the liquid in trying to stop the leak were burnt to death. This case illustrates the advisability of separation of any places where such accidental escapes may occur.

The first fatality caused by trichlorethylene was recorded in 1935. The tank in a metal degreasing plant was steamed out for a period of ten minutes after draining. Residue sludge in the tank still contained the solvent. A cleaner entered the tank assisted by an inexperienced sixteen-year old assistant. Both were found unconscious in the tank and the youth never regained consciousness. The total time of exposure was less than twelve minutes.

The more traditional danger from steam was emphasised again in 1936[100] by the explosion of an economiser, which caused the deaths of four persons and injured four others. A foreman caused the accident because he failed to close the flue damper after the water inlet and outlet valves had been closed. The economiser became a closed vessel in which steam was generated to a pressure beyond the safe limit of the economiser. Several explosions of compressed air receivers confirmed the requirement being set out in the new Factory Bill concerning testing, fitting of safety valves and periodic inspection. In the same year a dust explosion at a disintegrator used for grinding rice caused injuries to thirteen men and almost completely destroyed the factory. Flame or another heat source close to a vessel containing inflammable vapour was the cause of recurring accidents. This was the cause of an explosion at a tar distillation works where a worker attempted to weld the flange on a tar boiler. The resulting explosion killed four employees. Explosions at two gas works, which caused the deaths of four persons and injuries to two, were caused by persons in charge failing to ensure that the gas was isolated before the system was opened.

The enactment of The Factories Act 1937, which passed into law on 30 July of that year, had little immediate effect on the number of serious incidents causing multiple fatalities. A number of serious explosions were reported,[101] the worst being in a steel foundry where molten metal was explosively ejected while a 38-ton casting was being made. The molten metal escaped through a fissure in the mould and came in contact with the damp earth at the bottom of a pit. Eight workers received burns from which three died later. In a chemical works, much structural damage was caused and many were injured when an autoclave burst with tremendous violence.

The lid weighed about 12 tons and was hurled a distance of 700ft onto waste ground. Accidents caused by the application of heat during repair work on tanks, drums and closed vessels which containing petrol vapour, carbide and other flammable liquids continued to be a cause for concern. The new Act now required all practical steps to be taken to remove or render harmless any liquid or vapour remaining in the vessel. The additional danger of the widespread toxic risk was realised in 1938 [102] following the failure of an ammonia pressure vessel in an ice cream factory. A nozzle on an ice cream making machine was forced off and ammonia escaped into the factory to affect eighty-one of the 150 workers. The main effects on the workers were inflammation of the respiratory passages and burns of varying severity. Hospital treatment was needed in sixty cases and two subsequently died. A similar leakage from an ammonia compressor in a cooling plant in the yeast room of a brewery affected two men, one of whom died from his injuries.

Mr McNair retired as an Engineering Inspector from the Factory Inspectorate in 1938. Despite his important service to health and safety in the manufacturing and chemical industries, he had not progressed to a higher grade within the Department. One can only speculate that his contribution to the work of the Department and his dedication to solving the safety problems of industry provided him with a personal satisfaction that did not require the added status of promotion. The following year saw the retirement of Mr G. Stevenson Taylor OBE, who had also made a great contribution to safety in these industries. Unlike Mr McNair, he had received the rewards of promotion to Deputy Chief Inspector and public recognition for his work. It can be said however, that neither they nor their colleagues have been given full credit by the Department for their work over two decades when British industry was making the difficult transition towards modernisation.

Perhaps the greatest catastrophe to affect industry at that time occurred in Huddersfield during the early years of the war in October 1941. A clothing factory was burnt out in such a short space of time that forty-eight workers were killed and three were injured. The deaths included eleven men, nineteen women, one boy and seventeen young girls less than eighteen years of age. The fire started in a cloakroom near the main door on the ground floor. It started in such a small way that the first people to notice it thought they could get it under control and did not sound the alarm. When the alarm buttons were pressed, they failed to operate because the electric wiring had burnt through. It was a windy day, the rooms were wooden lined – a combination which caused the fire to spread rapidly. Fire drills previously had shown that the factory could be emptied in five minutes. Unfortunately, the alarms on the day did not reach the responsible persons in good time.

The Engineering Branch continued to provide the specialist skills for both the manufacturing industry and the chemical industry throughout the war. Mr F.E. Pollard, the Senior Engineering Inspector, was killed in a street accident in 1940 and was succeeded by Mr L.N. Duguid. 1944 saw the retirement of Mr C.W. Price, Senior Engineering Inspector, who, along with Mr McNair, was one of the first Engineering Inspectors when the Branch was formed in 1921. On 1 October 1944, the Engineering Branch became the Engineering and Chemical Branch with Mr L.N. Duguid as the Senior Engineering Inspector and the appointment of Mr S.H. Wilkes as the Senior Chemical Inspector. It had taken a long time for the Factory Department to recognise the need for a separate Chemical Engineering

Mr L.N. Duguid

Branch within its technical structure.

Reporting dangerous occurrences

The recognition of Dangerous Trades under the provisions for Special Rules and Requirements under the 1891 Act reflected the determination of Parliament to keep health and safety legislation in line with and relevant to the technological changes taking place in industry. Industrial accidents and the potential for accidents were more complex and not solely related to machinery. There was a need to have some means of recording incidents of a potentially serious nature so that suitable precautions were taken ahead of a more serious set of circumstances. The 1901 Act, which consolidated the legislation, also led to the making of the Notice of Accidents Act 1906. This Act required factory owners to report all accidents that came under a certain classification. Special Order changed the classification of dangerous occurrences from time to time thus making it difficult to identify trends.

Table 8.1 lists the number of reported accidents (including fatalities) under the heading of explosion, gassing and fire (not a dangerous occurrence) and it lists the number of incidents that were reported as dangerous occurrences under the heading of fire, for the period 1920 to 1935. Table 8.2 continues the figures from 1936 to 1958 except that the classification of dangerous occurrences is extended to cover six individual categories, which are listed after the table.

The Factory Inspectorate never had great faith in the accuracy and hence the value of recording dangerous occurrences, yet the statistics did provide valuable indicators to the likely source of accidents in industry. The frequent changes in the classification only helped to obscure these indicators. The 1933 Annual Report in commenting upon the significant increase in the number of reported dangerous occurrences from 2,656 in 1932 to 4,930 in 1933 concluded that this was because employers had been reminded of their obligations to report all chain breakage as dangerous occurrences. The report stated:

Little significance however, need be attached to this increase (which in any event is probably

Year	Reported Accidents						Dangerous Occurrences
	Explosions		Fires (not D.O's)		Gassing		Lifting appliances Fires Bursting vessels Etc.
	Fatal	Total	Fatal	Total	Fatal	Total	
1920	76	920	-	-	41	1032	2019
1921	40	594	-	-	15	115	1423
1922	33	694	-	-	20	193	1311
1923	27	726	-	-	15	280	2226
1924	60	848	-	-	21	218	1938
1925	33	1203	-	-	19	200	1504
1928	20	966	18	260	19	167	1628
1929	25	1071	4	266	15	212	2218
1930	50	934	10	245	17	133	2452
1931	33	751	11	221	11	122	2506
1932	15	693	5	205	10	168	2656
1933	17	761	11	194	14	140	4532
1934	16	818	8	256	21	176	5468
1935	23	905	19	280	13	120	5681

TABLE 8.1: Summary of accidents and dangerous occurrences from 1920 to 1935

Year	Reported Accidents						Dangerous Occurrences					
	Explosions		Fires (not DO's)		Gassing		Classifications (see below)					
	Fatal	Total	Fatal	Total	Fatal	Total	I	II	III	IV	V	VI
1936	34	1018	10	325	12	153	184	278	44	127	296	15
1937	25	1052	14	376	20	196	135	173	39	100	332	22
1938	24	1103	11	340	27	190	135	161	39	87	265	35
1939	37	1150	16	423	11	184	115	118	52	122	218	36
1945	51	1549		641	28	425	116	154	70	69	225	46
1946	33	1464	15	609	13	242	97	149	78	46	190	35
1947	24	1338	13	492	22	244	123	227	83	84	220	50
1948	55	1425	7	436	14	229	144	227	66	69	210	43
1949	36	1305	11	425	24	218	117	222	83	119	223	69
1950	34	1203	7	431	29	241	112	205	104	102	207	46
1951	41	1149	12	422	17	228	132	220	96	94	188	35
1952	13	1043	12	431	17	267	132	206	107	130	227	21
1953	27	1094	19	477	23	256	117	224	89	104	261	28
1954	27	1109	15	425	21	237	147	232	92	113	231	27
1955	49	1205	20	525	11	255	110	253	118	144	264	21
1956	27	1031	21	408	18	236	102	238	80	145	259	12
1957	27	1034	11	347	22	232	247	370	172	160	290	45
1958	23	978	12	349	12	218	220	372	151	140	327	16

TABLE 8.2: Summary of accidents and dangerous occurrences from 1936 to 1958

fictitious), since experience has shown that such chain breakage are in fact very rarely accompanied by accidents. [97]

The six dangerous occurrence categories I to VI were defined as follows:

I. Bursting of revolving vessel, wheel, grindstone or winding wheel moved by mechanical power.

II. Collapse or failure of a crane, derrick, winch, hoist or other appliance used in raising or lowering persons or goods (except breakage of a chain or rope sling) or the overturning of a crane.

III. Explosion or fire causing damage to the structure of any room or place in which persons are employed or to any machine or plant contained therein, and resulting in the complete suspension of work for not less than five hours where such explosion or fire is due to:

i) ignition of dust, gas or vapour

ii) ignition of celluloid

IV. Electrical short circuit or failure of electrical machinery, plant or apparatus attended by fire or explosion or causing structural damage thereto and involving its stoppage or disuse for not less than 5 hours.

V. Explosion or fire affecting any room in which persons are employed and causing complete suspension of ordinary work therein for not less than 24 hours.

VI. Explosion of a receiver or container used for the storage at a pressure greater than atmospheric pressure any gas or gases (including air) or any liquid or solid resulting from the compression of gas.

CHAPTER 9

THE ELECTRICAL ENGINEERING BRANCH

Discovery and development of electricity

Faraday's discovery of the principle of electro-magnetic induction was made about the same time as the creation of the Factory Department. Arc lamps were used in the Paris Opera House in 1844. The early forms of arc light were superseded by the incandescent lamp and were demonstrated at the Electrical Exhibition held at the Crystal Palace in 1881. It was reported in the journal *The Engineer* [48] of 30 September 1881:

> William Cammell & Co. at Drifield Steelworks near Sheffield have established the utility of electric light for the illumination of large steelworks both as regards to its effulgence and on grounds of economy. This first important industrial establishment in the district would be watched with interest and will probably lead to its being brought into requisition in other large steel and iron works.

Electricity was also making some headway in public lighting and *The Engineer* observed the following:

> Godalming is one of the first small towns to use an available water supply to provide streets and public buildings with light instead of burning coal brought from a considerable distance. Experiments were carried out using a water-wheel on the River Wey. One large Siemens arc lamp and several of Swans small incandescent lamps produced illumination of a highly satisfactory character.

It was soon apparent to the fire insurance companies that a new risk needed attention. In 1882, Mr Musgrave Heaphy, the engineer to the Phoenix Fire Office, drafted rules which were published, and it is believed represented the first electrical regulations designed for the purpose of safety. They were well received by the principal electric light companies and were followed by a code prepared by the Society of Telegraph Engineers, the predecessors of the Institution of Electrical

Engineers, in which appeared the first provisions against the danger from electric shock. The Society considered that no one should be exposed to shocks from alternating current at pressures in excess of 60 volts.

The Engineer of 20 May 1892[48] reported on the Crystal Palace Exhibition where Sebastian de Ferranti exhibited new alternating current motors which were non synchronous, could start with a load and could be stopped or varied in speed with ease. They had no commutators and were self regulating. Ferranti was reported as claiming that the AC system was better adapted for motive power than those of continuous current. Sometime later the journal had this to say about these developments:

> We are of the opinion that the day is not far distant in which electricity will become a powerful rival to belt, wire rope and cotton rope transmissions to diminish losses . . . Whether in the case of large machine tools it would be better to discard shafting and belts all together and supply a special motor to each tool is a question which must be settled in each individual case.

Mr G. Scott Ram

The Board of Trade became officially concerned with electrical safety in 1882 but it was not until 1896 that a Departmental Committee of the Home Office looked into the conditions of work in electrical generating works. There was very little activity in the department thereafter until the passing of the 1901 Act, which included Electrical Stations. Mr G. Scott Ram was appointed as the first HM Electrical Inspector of Factories to administer these new provisions.

In 1907 the Secretary of State certified the generation, transformation, distribution and use of electricity in factories as dangerous and drafted regulations which were prepared by Mr Ram. These regulations were necessary because of the number of workshops being changed over to factories because of the increasing availability of electricity and the facility with which electrically driven manufacturing machinery could be installed. Consultation on the draft regulations took place and points of dispute were referred to a technical committee composed of the most competent engineers of the day. The code that emerged was not materially different from the original written by Mr Scott Ram. Many years later it was written: [96]

> This high tribute to the mind which conceived the Electricity Regulations has been justified in subsequent administrative experience, and it is a remarkable fact that the code has, without revision, been successfully applied to an industry in which technique has changed out of all recognition and which has grown to dimensions which must have exceeded the anticipation of the most far-sighted.

The Electrical Branch came into being with Mr Scott Ram's appointment in 1902. No further Electrical Inspector posts were created until 1921. During this critical period in the development of electrical power in industry, Mr Scott Ram provided the only specialist advice to the Department. The 1910 Annual Report[80] for example notes, 'Mr Scott Ram's time is fully occupied with the recently issued Electricity Regulations'. It also records that advances in the use of electrical energy show an additional 70,000 horsepower for electric motors connected to the public supply. Electricity was finding increasing use for electric furnaces and providing the motive power in the manufacture of steel. District Inspectors were reporting that gas engines were being displaced by electricity except in some large companies because of the economy of available producer gas.

Beginning of telecommunications

Alongside the development of electricity for power purposes, existed the equally rapid development in the application of electro-magnetic waves for wireless telegraphy and the later development of the radio and broadcasting for public entertainment. Guglielmo Marconi was one of the earliest pioneers in this field. He was born in 1874 in Bologna in Italy and by 1895 he had succeeded in transmitting signals over a few yards of space by means of two insulated plates separated by a spark gap consisting of two small spheres connected across the secondary of an inductance coil, the primary of which included a battery and Morse key. The Italian Government showed very little interest in his work so he emigrated to England in 1896 where he filed the world's first patent application for a system of telegraphy using Hertzian waves. In 1897, Marconi established communication across the Bristol Channel and set a new record for the transmission distance of 8.7 miles. On 12 December 1901, in St John's, Newfoundland, he picked up the three dots of the Morse code 'S' being transmitted from Poldhu in Cornwall, to make the first link across the Atlantic. 1901 also saw the first patent for radio valves, filed by Dr J.A. Fleming, Scientific Adviser to the Marconi Company. In 1910 a Marconi spark transmitter was used to send the first message from an aeroplane to a ground station. When the *Titanic* struck an iceberg in 1912, those who survived owed their lives to the wireless distress calls made before the ship sank.

Marconi founded the Wireless Telegraph and Signal Company and premises were set up in Chelmsford in Essex. This company became the Marconi Wireless Telegraph Company in 1900 and the works in Chelmsford became the first radio factory in the world. Wireless broadcasting as we know it started in Chelmsford in 1919 in the form of an experimental transmitter. On 15 June 1920 in Britain's first advertised public broadcast programme, Dame Nellie Melba broadcast a song recital from the Marconi works. The Postmaster General granted the company a licence allowing for regular, although very restricted, broadcasting and transmission started in 1922. A further licence was granted for the 2LO station in Marconi House in London. Enthusiasm of the public for 'listening in' was such that the Postmaster General had to call a conference of manufacturers. It was decided that the six largest companies, including Marconi, should form the British Broadcasting Company, which in 1926 became the British Broadcasting Corporation.[53]

This rapid development in the new technology of electrical power and communication led to concern about safety and reinforced the need for legislation that was in keeping with the pace of change. In 1910 there were 276 electrical accidents including 5 fatalities. A Memorandum on the Electricity Regulations was written by Mr Scott Ram and published at the price of 3*d*. He had this to say about the regulations in the Annual Report[80] for the year:

> With regard to observance of the regulations . . . vast numbers although making use of electrical energy understand little or nothing about it and are quite incapable of determining whether the regulations are complied with or not . . . Requirements were drawn up so as to give the utmost latitude to design of apparatus and methods of protection so as not to check or retard progress, provided that the object — safety to the user, and other persons employed, is attained . . . Reports from the Inspectors show a good deal of work has been done in getting some of the more obvious defects remedied such as bare connectors, fuses and switches etc. repaired.

The Electrical Engineering Branch at work

The Electrical Branch worked closely with other government departments and with the electrical industry during this critical period when new and unfamiliar equipment came into use. The extent of this involvement is illustrated in the 1920 report[86] which stated:

Electrical manufacturers have done much to improve apparatus and accessories from the safety point of view, and not infrequently describe apparatus in their catalogues of advertisements as being of 'Home Office pattern' or as 'complying with the Factory Act requirements'. In many cases, particularly as regards hand lamps and fuses, the apparatus in question has been submitted for comment, and in some cases clearly does not comply with the requirements. In a recent case a hand lamp, so described in a trade catalogue, was found to be wrongly constructed in all essential particulars and was only discovered as a result of a fatal accident occurring through its use.

Mr Scott Ram continued as the only Engineering Inspector until 1921 when four new Electrical Inspectors were appointed. Although the need for more Electrical Inspectors was recognised as far back as 1914, the intervention of the War had prevented the appointments being made. The new Inspectors were recruited from outside the Department. Developments in electrical engineering meant that only trained and experienced electrical engineers would have the necessary degree of technical knowledge to deal with the many aspects of the work. The Senior Electrical Inspector's Report for 1921[87] was Scott Ram's twentieth and it gave him the opportunity to consider the progress made since his appointment in 1902. The addition to his staff was clearly necessary in order to be able to effectively inspect the 600 electricity supply undertakings containing generating stations and several thousand substations. There were 70,000 works registered where electricity was used for driving machinery and for other purposes. He gave the following account of the developments over the last twenty years on the generation and use of electricity:

> The broad features, briefly enumerated, are, as regards generation and distribution, the gradual disappearance of the Lancashire boiler in favour of the water tube types, the adoption of the steam turbine of large size in place of the reciprocating engine, the generation and distribution of three-phase current at high pressure in place of direct current at low pressure or single-phase alternating.
>
> On factory premises in 1902 the use of electricity for driving machinery was only beginning to make headway. Direct current at low or medium pressure was mostly used and the motors were of comparatively small size. Now three-phase current is the more widely adopted, and large motors are connected directly to the high-tension supplies. The generation and use of electrical energy has been on the increase throughout the period, with a very large impetus during the period of the war, when the advantages of electrical driving were forcibly brought home to manufacturers.
>
> Numerous applications of electrical energy have been developed and put into use on factory premises, e.g., the driving of heavy rolling mills by electric motors, electric furnaces for the manufacture of high grade steel, electric welding both by resistance and arc methods, many electro chemical processes notably for the manufacture of aluminium. The electrification of suburban railways has also made great progress, only three lines being worked in 1902. The generating and substations of these railways are subject to inspection by the Department.

There were 322 accidents including twelve fatalities in 1921. Twenty-one accidents occurred on high pressure or extra high-pressure systems and seven of these were fatal. Seventeen accidents including five fatalities occurred in public supply stations. A large proportion of the accidents (70) on the low and medium pressure systems were due to persons cleaning or repairing live switchboards or other apparatus. A pamphlet, 'Memorandum on Electric Arc Welding', Form 329, dealing with the precautions necessary in respect of dangers arising from this new process was issued by the Department and proved to be very useful. Portable apparatus and connectors accounted for fifty-seven accidents and a number occurred in the renewing of fuses not constructed or protected as required by the regulations. Table 9.1 contains a summary of the electrical accidents reported under the Factories Act and also reports of fatalities in premises outside the Act for the period from 1907 to 1958.

Year	Reported under Factories Act		Accidents outside Factories Act		Year	Reported under Factories Act		Accidents outside Factories Act	
	Total	Fatal	Total	Fatal		Total	Fatal	Total	Fatal
1907	323	9			1939	589	29		
1908	309	12			1940	729	32	118	
1909	318	13			1941	921	51	119	
1910	327	9			1942	1042	51	130	
1920	403	23			1943	1255	58	107	
1921	322	12			1944	1072	31	123	
1922	309	17			1945	937	31	147	
1924	453	37			1946	769	33	113	
1925	414	24			1947	734	30	95	
1928	427	26			1948	780	43	104	
1929	420	35			1949	771	24	105	
1930	359	21			1950	778	38	86	
1931	321	22			1951	715	34	73	
1932	313	19			1952	721	38	98	
1933	346	25			1953	744	40	86	
1934	380	31	50		1954	707	33	89	
1935	447	23	64		1955	739	42	100	
1936	520	31	81		1956	797	40	104	
1937	583	36	70		1957	687	32		
1938	560	30	97		1958	714	38		

TABLE 9.1: Electrical accidents from 1907 to 1958.

The Senior Electrical Inspector's report for 1924[90] noted the rapid development and activity of the electrical supply industry. The old parochial boundaries of electricity supply were swept away and many stations were extending their operations. Transmission voltages of 6,000 volts and 11,000 volts, which at one time were very high, were now common and there were a number at 33,000 volts and one operating at double this voltage. He observed that on all new systems the electrical current was delivered as AC and the DC in many older systems was being changed over. The number of transforming and distributing sub-stations had greatly increased. Some twenty years after the Electricity Regulations were in force, Mr Scott Ram was able to report[92] that although enormous strides had been made in the design of electrical plant and the methods of supply and in new applications of electricity, the regulations continued to be relevant.

A Home Office Departmental Committee reported in 1928 on the Electrical Branch, which at that time comprised one Senior Engineering Inspector at Headquarters and four Engineering Inspectors out-stationed at London, Birmingham, Leeds and Glasgow. They were responsible for the inspection of works that came under the Electricity Regulations, now applying to over 100,000 premises and 13,000 sub-stations. 17,000 premises required specialist inspection, including 570 generation stations, railway companies and tramway authorities and over 2,700 other generation stations as sub-stations under the Factories Act. The Committee noted that Mr Scott Ram was kept in his office dealing with reports and other matters. They felt that it was important that he spent part of his time in active inspection of electrical installations to keep in touch with important developments. They recommended the appointment of a deputy to assist him. They also concluded that the present salary (£400 by £20 increments to £650) was not sufficient for recruitment even with a cost of living bonus. They felt that suitable candidates for Electrical Inspector would be more experienced and older than general staff. This would affect their pension and promotion prospects outside the Branch; therefore the Committee recommended a salary increase because the present scale was inadequate. This differential in salary between the Electrical Inspectors and other specialist inspectors remained in place up to present times.

Two years later on 1 December 1930, Mr Scott Ram retired after twenty-nine years' service. Mr H.W. Swann superseded Mr Scott Ram as the next Senior Electrical Inspector of Factories. For the greatest part of his service, Mr Scott Ram had worked single-handed. He had drafted the Electricity Regulations at a time far removed from subsequent developments yet they had remained a valid means of ensuring the safety of those who through their work came into contact with all types of electrical plant and equipment.

An analysis of the apparatus causing accidents was presented in the 1930 report,[94] in a year when there were a total of 359 reportable accidents including twenty-one fatalities. This is summarised in Table 9.2 which shows that switchgear operating below 650V was responsible for the largest number of accidents. This covered switchboards, fusegear and motor control apparatus, which generally caused burns following a short circuit. Failure of the insulation on cables and flexibles led to many cases of severe burning. The report, although written more than seventy years ago, retains a relevance to today's experience:

> Flexibles are usually left too long in service and although extensive use is made of tough rubber sheathing the failures occur at entries to fittings mainly owing to misuse of, or absence of effective sheath grips. It is evident that present-day conditions are tending to a definite increase in the severity of short circuits and this fact in conjunction with over-fusing of circuits, is accountable for a good deal of the trouble.

The common use of screw cap-type lamps at that time was responsible for six of the reportable fatal accidents listed in Table 9.2. These accidents were caused by exposure of the live caps in unskirted holders. The following warning was given in the report:

Apparatus	Fatal	Non-fatal		Total
	Men	Men	Women	
Electrical machinery	-	9	-	9
Transformers	3	3	-	6
Switchgear above 650 volts	-	22	-	22
Switchgear below 650 Volts	1	107	2	110
Crane trolleys	3	16	-	19
Electric welding	-	14	-	14
Portable electric machines	-	15	-	15
Portable heaters and irons	-	3	9	12
Portable lamps	6	15	-	21
Fixed lamps	3	16	2	21
Cables and flexibles	2	60	2	64
Electrical ignition of inflammable materials	1	2	1	4
Testing	1	34	2	37
Other apparatus	1	4	-	5
TOTAL	21	320	18	359

TABLE 9.2. Electrical accidents in relation to apparatus - 1930

There are four ways of wiring a simple circuit comprising a single pole switch controlling a screw cap lamp and of these three combinations are incorrect. Mistakes are therefore to be expected and in addition lighting circuits with single pole switches are sometimes connected between phases of three phase systems, both conductors being live. There is therefore, a strong case for the use of holders with skirts long enough to prevent the lamp screw from being touched so long as it remains in engagement with the holder. The point has been discussed with the leading electrical manufacturers concerned with the production of screw cap lamp holders and most of them are now marketing fully protected holders. [94]

A notable feature of the statistics for electrical accidents, as shown in Table 9.1, was that the numbers remained relatively unchanged over a period of fifty years despite the rapid technological changes that took place. To some extent the figures were misleading because the statistics were limited to reported accidents in factories – there were many more accidents in places outside the jurisdiction of the Factories Act, also shown in Table 9.1. Nevertheless, there was no evidence of the type of variation that can be seen in other industries, caused either by uncontrolled dangers or fluctuations in the numbers employed. It was reported[97] that over the ten-year period from 1923 to 1933 there was an increase of 132 per cent in the number of units of electricity sold by Authorised Undertakings and the increase in factories under the regulations rose from 78,820 in 1923 to 127,756 in 1933 which was the year that saw the completion of the national grid system by the Central Electricity Board. In a year of great importance to the electricity industry the report was able to conclude:

> The Regulations continue to influence the design of apparatus and methods of installation, and in many directions improvements have been achieved. Close contact with electrical manufacturing interests is maintained by inspection and interviews, and it is pleasing to

Illustrations of the type of accidents caused by the apparatus listed in Table 9.2.

record the willing co-operation, which is almost always forthcoming. The Electrical Branch has been represented on various committees dealing with many aspects of electrical work in nearly all of which questions of safety arise, experience is showing the value of work of this character.

With the increasing threat of war in 1939, the Electrical Branch in addition to ordinary duties undertook special work on lighting problems in connection with outside work on docks, shipyards and buildings under construction. The Senior Electrical Inspector, Mr Swann and his staff advised various government departments on electrical safety problems connected with the war effort. In 1940[104] the Electrical Branch was unable to give a detailed report because the work was of a kind that could not be widely disclosed in wartime. They were able to say however that much of their work was of such great interest that the time would soon come when an account could be published. The first evidence of that account came in 1944,[108] which noted that the war years had seen many new applications of electricity in factories. The degree of risk presented by these developments and the methods of dealing with them had to be assessed and devised under the inevitable pressures of the war effort. Some of the most difficult problems that had to be dealt with were in connection with the aircraft industry where the use and testing of large volumes of aviation fuels presented a large element of danger. Special legislation was brought in to deal with these risks in the form of the Factories (Testing of Aircraft Engines, Carburettors and Other Accessories) Order 1944, which was largely the codification of a series of recommendations made by the Electrical Branch during the war years.

Additionally, the Electrical Inspectors had undertaken considerable executive and advisory work for the Ministry of Home Security, including advise on the Lighting (Restrictions) Order and its application to shipyards and ports. From 1940 onwards, the Inspectors advised all the principal seaports on ways and means of making the maximum use of permitted standards of lighting and to control this in relation to the reception of air-raid warnings. The Annual Report illustrated the extent of their responsibilities in this critical part of the war effort:

> Whilst the earlier work was advisory, that undertaken more recently in connection with the 'operational ports' was executive and the Inspectors had to find both material and labour for the lighting installations at a time when everything was in short supply. Their close connection with the Electrical Industry proved of great value in getting round, over and through this trouble and they received the utmost assistance especially from the Electrical Supply Undertakings along the coast and elsewhere. The work has brought out the fact that a large number of lights of low intensity compare well for the purposes of work with a small number of lights of high intensity; in fact the Engineers at a good number of ports have expressed the opinion that their premises were now better lit than at any previous time. There is no doubt that this work has contributed greatly towards the reduction of drowning at night in the darkened ports and also to the reduction of other accidents.

The common safety objective being followed by Electrical and Mechanical Engineering Inspectors was confirmed at the end of the war when the Senior Electrical Inspector recorded[112] that his Inspectors made important contributions to the safety of operations that were not essentially electrical and important advisory work had been done on the safe control of machinery. He concluded that attention must be paid to sound mechanical design of the electrical components and a proper study of electrical circuits to avoid danger due to purely electrical defects and to ensure 'failure to safety'. If electrical features of the equipment were to be an effective part of the completed design or installation, then in his opinion, 'electrical and mechanical (and indeed chemical) features must be considered together and coordinated by the draughtsman, and in the design office as to ensure a well integrated whole'.

CHAPTER 10

THE MEDICAL BRANCH

Role of the Certifying Surgeons

The Act of 1833 required the Factory Inspectors to make such rules as might be necessary for the enforcement of the Act. Each Inspector made detailed orders for his own District on such matters as the keeping of registers and time books. Local Surgeons were appointed for the sole purpose of granting certificates of age for young persons. The hours worked by young persons and children, though limited were not fixed, therefore the legality of their appointment could only be proved by ascertaining the age of the child. The Inspector had to depend upon the accuracy of the Surgeon's certificate. The children and their parents were understandably disinclined to admit illegal employment because in many cases the child was a main provider to the household. Records show that some Surgeons took their duties lightly and others deliberately provided false certificates. Leonard Horner reported on this subject in 1838:

> I did discover some instances of gross negligence on the part of Surgeons and I cancelled the appointment of two; the one who had been detected in a very discreditable transaction with regard to a certificate; the other who had acted so culpably that my Superintendent very properly laid an Information against him and he narrowly escaped imprisonment for issuing false certificates.[96]

The Report of the Factories Inquiry Commission of 1833 had clearly identified the health problems being experienced by the factory workers:

> That excessive fatigue, privation of sleep, pain in various parts of the body, and swelling of the feet experienced by young workers, coupled with the constant standing, the peculiar attitude of the body, and the peculiar motion of the limbs required in the labour of the factory, together with the elevated temperature and the impure atmosphere in which that labour is often carried on, do sometimes ultimately terminate in the production of serious, permanent, and incurable diseases, appears to us to be established.

The regular protests by the Inspectors about dirty conditions in the factories and their effect on health were eventually addressed in the Act of 1844. This required the periodic lime washing of inside walls and ceilings, passages and staircases in a factory unless painted with oil paint, and the washing of oil painted walls and ceilings. Under the 1844 Act, power was given to the Inspectors to appoint a sufficient number of persons practicing surgery or medicine to be Certifying Surgeons for the purposes of examining persons brought before them to obtain surgical certificates of age. By the same Act it became a statutory duty for the Certifying Surgeon to investigate and report on accidents occurring in any factory for which he had been appointed to grant certificates of age. But it took a long time before the professional skills of the Certifying Surgeons and the science of medicine was considered necessary as an aid to improving the working conditions in factories.

Concern about substances injurious to health

Wet spinning of flax was of particular concern because the process was carried out with hot water without any protection being provided for the young people employed there. Their health was at great risk from the effects of being confined in rooms filled with steam and being constantly sprayed with hot water. In consequence, women, young persons and children were prohibited by the 1844 Act from employment in flax spinning unless sufficient means were provided to protect the workers from being wetted and to prevent the escape of steam into the room. In 1861 the Medical Officer of the Privy Council carried out inquiries into the death rates from lung diseases in certain industrial occupations. This resulted in the Factory Act of 1864, which required that every factory should be ventilated to render harmless, so far as was practicable, any gases, dust or other impurities that might be injurious to health. The Medical Officers report contained the following statement:

> Unboxed machinery against which he (Factory Inspector) now has authority to move the magistrate to penalties is indeed a danger to life and limb; but even though every mutilation which results from it were to be counted as a death, the deaths from unboxed machinery would probably count as nothing in comparison with those which the unventilatedness of factories occasions...The canker of industrial diseases gnaws at the very root of our national strength.[96]

Phosphorus poisoning came to their attention in 1880 in Oldbury where seven cases were recorded over a period of twenty-nine years. In one instance the disease had resulted in the loss of a lower jaw, leading to the description of 'fossy jaw'. Anthrax attracted concern at the same time, when several deaths were reported in Bradford among workers involved in the sorting of foreign wool and hair (anthrax was often called 'woolsorters' diease'), which led Dr Bell to identify the bacterium *Bacillus anthracis* as the cause of the disease, and he published his conclusion in 1876. At a lecture before the Society of Arts in 1884, Dr B.W. Richards made reference to a disease affecting flour millers caused by dust. In Sheffield in 1887, attention was drawn to an excessive high death rate from phthisis and other respiratory diseases among grinders, cutlers, tool, fork and scissors makers. Flint dust in potteries also caused concern because of the absence of power in many potteries to drive mechanical exhaust ventilation fans. Exhaust ventilation by the means of fans, which was required by the 1878 Act, led to the development of new devices to improve that obtained from natural ventilation from doors and windows.

Considerable attention was directed in 1884 to ventilation by means other than windows and doors, such as by Tobin's tubes, louvres, cowls, Boyles air pump ventilator, the 'hit and miss' and the like. It is said 'In such places as the rag house of paper mills the Blackman air propeller, if judiciously fitted, does all that is wanted. Where there is much dust its merits are conspicuous'.[96]

Section 36 of the 1878 Act required exhaust ventilation by means of a fan to be provided for the removal of dust likely to be injurious to health. Inspectors were empowered to obtain better sanitary conditions in factories. Children and young persons were prohibited from employment in the manufacture of white lead and the silvering of mirrors with mercury; also in the processes of dry grinding of metals and the dipping of lucifer matches except when made with red amorphous phosphorus. In certain factories, including white lead works, women and young persons were prohibited from taking meals in rooms in which a process was carried on giving rise to dust. The 1891 Act was the most important in this respect because it gave power to the Secretary of State to certify any machinery, process or particular description of manual labour used in a factory or workshop to be dangerous or injurious to health. By this means the Chief Inspector was enabled to serve on the occupier of the factory or workshop, a notice in writing, proposing special rules or requiring the adoption of special measures as appeared to the Chief Inspector to be reasonably practicable. Typical of such special rules are those below which were made for Lucifer Match Factories where white or yellow phosphorus was used:

SPECIAL RULES for LUCIFER MATCH FACTORIES
Where White and Yellow Phosphorus is used

DUTIES OF OCCUPIERS

I. It shall not be lawful to carry on a lucifer match factory, where white or yellow phosphorus is used, unless such a factory is certified by an Inspector to be in conformity with the following special rules.

II. All occupiers of such factories shall provide for the processes of mixing, dipping and drying an apartment or apartments separate from other portions of the factory.

III. They shall take effective means to prevent the fumes from the before-mentioned processes and from the boxing department being allowed to enter the rest of the factory.

IV. They shall provide efficient means, both natural and mechanical, for the thorough ventilation in the mixing, dipping, drying and boxing departments.

V. They shall provide washing conveniences, fitted with a sufficient supply of hot and cold water, soap, nail brushes and towels, and shall take measures to secure that every worker washes his or her hands before meals, and before leaving the works. Managers and overseers shall report immediately to the occupier any instance which comes under their notice where this regulation has been neglected.

VI. Any person employed in the works complaining of toothache, or of swelling of the jaw, shall at once be examined by a medical man at the expense of the occupier; and if any symptoms of necrosis are present the case shall be immediately reported to one of Her Majesty's Inspectors of Factories for the district.

VII. No person having suffered from necrosis shall be permitted to resume work in a lucifer match factory until a certificate of fitness has been obtained from a qualified medical practitioner.

VIII. No person shall be permitted to work in the processes of mixing, dipping, drying or boxing after the extraction of a tooth, without the certificate of a duly qualified medical practitioner that the jaw has healed.

AS TO PERSONS EMPLOYED

IX. Every person employed in the mixing, dipping, drying or boxing departments shall carefully wash his or her hands and face before meals and before leaving the works.

X. In all cases where the co-operation of the workers is required for the carrying out of the foregoing rules, and where such co-operation is not given, the workers shall be held liable in accordance with the Factory and Workshop Act, 1891, Section 9, which runs as follows:-

" If any person who is bound to observe any special rules established for any factory or workshop under this Act acts in contravention of, or fails to comply with, any such special rule, he shall be liable on summary conviction to a fine not exceeding two pounds."

B.A. WHITELEGGE HM Chief Inspector of Factories

Following the passing of the 1891 Act, special enquiries were carried out in those industries where dangerous materials were handled and injurious effects were known. Special rules were immediately drafted to cover works handling white lead. This required the weekly examination of workers by the Certifying Surgeon and the keeping of a register. This was quickly followed by special rules for the manufacture of paints and colours, the extraction of arsenic, enamelling of iron plates, manufacture of lucifer matches, manufacture of explosives in which dinitro-benzol was used and for chemical works. In the following years the Dangerous Trades Committee appointed by the Secretary of State carried out Special inquiries. The Committee sat from 1893 to 1899 and inquired into many different industries and processes. As a result of these inquiries and the collective experience from previous years, certain industries and processes were identified and subject to the special rules and new regulations to protect those who were exposed to the dangerous materials.

Exposure to lead and other hazardous substances

Lead was probably the first hazardous material to attract the attention of the Factory Inspectors. In 1883, an Act was passed to regulate white lead works. Before a white lead factory could be permitted to operate, a certificate had to be obtained from an Inspector. It is recorded:

The Act followed representation to Mr Redgrave from Guardians of the Poor consequent upon the number of cases of lead poisoning seeking poor relief and admitted to the infirmaries. The number of cases of lead poisoning among persons working in white lead factories, requiring practical relief in the Poplar Union in 1881 was 26, of whom 21 were females. From a report of the Medical Superintendent of the Holborn Union Workhouse Infirmary made in June1882, it appears that during the previous twelve months, 54 patients were admitted suffering from lead poisoning − 6 males and 48 females. The males were not employed in white lead works, but as painters, type setters etc., while the females were all employed in white lead works.[96]

The seriousness of lead poisoning among women was emphasised by the report of one Inspector that recorded the case of a girl starting work at the age of fifteen and dying from lead poisoning at eighteen. Another source of lead poisoning occurred among file cutters in Sheffield, where it was revealed that of 6,760 patients admitted to the Sheffield Public Hospital, there were fifty-two cases of lead poisoning, forty-two of which were file cutters. The manufacture of earthenware was another dangerous trade where the use of lead for glazing and colour making was a cause of plumbism among the workers. Special rules for earthenware and china were drafted by the Chief Inspector in 1892, much to the disapproval of the manufacturers in the potteries trades. The code provided for the removal of dust from flint by fans or other means in the process of towing earthenware and the scouring of china. Overalls and head coverings were required in processes where exposure to lead was possible. Brushes had to be provided, and workshop floors had to be swept every day after sprinkling with water.

The Dangerous Trades Committee found that lead and all its compounds were to some degree dangerous to health. As a result, special rules were made covering other industries such as the manufacture of red, orange or yellow lead, lead smelting, the tinning of hollow-ware, electric accumulator works, brass mixing, casting, and the heading of yarn with yellow chromate of lead. Cases of ill health and death from lead poisoning were made notifiable by the Factory and Workshop Act of 1895. But it took a period of fifty-five years before the annual number of deaths was reduced to naught. This trend of notified cases of lead poisoning from 1900 to 1958 is shown in Table 10.1.

The 1895 Act required every medical practitioner to notify to the Chief Inspector, cases of mercurial and arsenical poisoning. Arsenical poisoning was caused by exposure to materials used by colour printers, lithographers and wallpaper manufacturers. Cases of phosphorus poisoning became notifiable because of its occurrence in the lucifer matchmaking industries. Anthrax was another illness that caused concern and was found in those industries where foreign wool and hair were sorted. As industries developed during this period, other materials came into use in the new manufacturing processes. New industrial illnesses became apparent and were added to the list of notifiable cases of poisoning recorded in the Annual Reports of the Chief Inspector. These included illness from exposure to the fumes of carbon bisulphide in the vulcanising of rubber. Epithellomatous ulceration was a particularly dangerous illness that exacted a high price upon those exposed to the deleterious effects of pitch, tar, paraffin and oil. Chrome ulceration was another

Year	Lead		Mercury		Anthrax		Epith. Ulceration		Chrome Ulceration		Aniline		Dermatitis	
	Total	Fatal	Total	Fatal	Total	Fatal	Total	Fatal	Total	Fatal	Total	Fatal	Total	Fatal
1900	1058	38	9	0	-	-	-	-	-	-	-	-	-	-
1910	505	38	10	1	51	9	-	-	-	-	-	-	-	-
1919	207	26	7	0	57	9	-	-	-	-	-	-	-	-
1920	243	23	5	0	48	11	45	1	126	0	-	-	-	-
1923	337	25	4	0	46	5	58	4	58	0	-	-	-	-
1924	486	32	5	0	43	4	123	24	45	0	-	-	-	-
1926	332	46	4	1	38	3	187	49	55	0	33	1	-	-
1927	347	35	4	1	38	3	187	49	55	0	38	1	-	-
1928	326	43	4	0	45	8	175	59	70	0	41	0	-	-
1929	244	31	0	0	40	5	165	50	109	0	26	0	-	-
1930	265	32	0	0	40	5	165	50	109	0	24	0	-	-
1935	168	17	1	0	20	3	171	33	67	0	-	-	-	-
1936	163	13	-	-	30	1	142	27	84	0	7	1	-	-
1937	141	19	7	0	23	4	183	31	101	0	-	--	1985	0
1938	96	19	2	0	34	5	165	21	115	0	9	0	2195	0
1939	109	6	10	0	37	5	160	31	159	0	12	0	2952	0
1940	108	6	5	0	37	5	166	15	121		64	0	4744	0
1941	59	5	5	0	22	3	128	11	103	0	249	0	7291	0
1942	72	8	1	0	25	1	113	8	89	0	204	0	8802	0
1943	46	5	4	0	17	4	160	15	226	0	79	0	8926	0
1944	41	5	7	0	8	3	205	20	121	0	55	0	8180	0
1945	45	2	5	1	7	0	215	9	94	0	31	0	5996	0
1946	47	8	1	0	14	1	245	32	96	0	19	1	6166	0
1947	58	2	1	0	25	2	203	16	296	0	16	0	6166	0
1948	49	2	2	0	32	0	233	18	146	0	12	0	-	
1949	53	2	0	0	2	1	190	13	139	0	12	0	3533	0
1950	57	0	0	0	36	0	195	13	143	0	6	0	3571	0
1951	64	0	3	1	31	1	178	1	203	0	5	0	3281	0
1952	48	0	8	0	20	1	157	2	217	0	12	0	3122	0
1953	52	0	0	0	29	1	256	68	164	0	15	0	3121	0
1954	49	0	1	0	18	1	173	12	220	0	11	0	2930	0
1955	69	0	2	0	15	0	211	18	261	0	9	0	2902	0
1956	49	1	2	0	19	3	199	23	189	0	19	1	2472	0
1957	55	0	12	0	14	1	197	14	213	0	16	0	2093	0
1958	55	0	4	0	6	0	176	16	205	0	7	0	1791	0

TABLE 10.1: Notifiable cases of industrial poisoning from 1900 to 1958.

illness that was experienced in the manufacture of bichromate, dyeing and finishing, chrome tanning and other industries. In time, other materials and their associated illness were added, including aniline poisoning, toxic jaundice, compressed air sickness and finally dermatitis, which became prevalent during the Second World War. A summary of the more serious cases of industrial poisoning between 1900 and 1958 is given in Table 10.1 above.

Appointment of the first Medical Inspectors

The Cotton Cloth Factories Act 1889 came into force to combat the health problems experienced by workers who were exposed to the effects of artificial moisture, used to control excessive dryness in the material, in the cotton works. This caused the workers to suffer from rheumatism, chest complaints and coughs. Mr E.H. Osborn, who was a Factory Inspector in the Rochdale Area, was charged with the enforcement of the new Act, which provided schedules of permitted temperatures and maximum humidity. The Factories Act of 1895 extended the provisions to every textile factory. A new Cotton Cloth Factories Act of 1897 was eventually superseded by the 1901 Act. In 1899, the specialist nature of Mr Osborn's duties was recognised by his appointment as the first Engineering Adviser.

The increasing complexity of industry and the growing awareness of occupational ill health of workers highlighted the need for other specialist skills within the Factory Department. This was first recognised by the appointment in 1896 of Dr B.A. Whitelegge as the Chief Inspector of Factories and Workshops. It was an important stage in acknowledging the importance of medical involvement in the work of the Inspectorate. Two years later this led to the appointment of Dr Thomas Legge as the first Medical Inspector at a time when other health problems were being experienced in the earthenware and china industries. In 1910, a second Medical Inspector, Dr E.L. Collis was appointed, having been previously employed as a Certifying Surgeon for Stowbridge and South Stafford. An historic review of the work of the Medical Department by Dr J.C. Bridge, gives the following account:

> During this period the Department made inquiries from the medical point of view into industries under special rules, in particular potteries and file cutting. Mercury poisoning among hatter's furriers, and in the making of philosophical instruments, and gassing by carbon monoxide and hydrogen sulphide, were also subject of inquiry. The chemical industry was evidently becoming alive to the risk of gassing, and an account of a patent safety respirator in use at the United Alkali Works at Widnes is given. This apparatus, which was on the lines of a smoke helmet with air pipe and hand bellows attachment, was used for rescue work. [96]

After the passing of the Workman's Compensation Act 1906, the work of the Medical Inspectors included consideration of diseases that could be made the subject of compensation. The Senior Medical Inspector served on various committees which inquired into these diseases and advised the Secretary of State on the need to bring new diseases, such as that caused by tetrachlorethane, under the Act. Following the appointment of Dr Collis, the Medical Department extended its field of interest and began a study into the question of injury of health from dust; 'which, while not producing illness or death by reason of its toxic character, did produce illness and death by reason of its effect on the lungs'. Dr Collis commenced with a study of pneumoconiosis in a number of industries:

> The period from 1900 to 1913 was, therefore, one of the close study of industrial diseases in which the medical and other technical branches of the Department were becoming more closely associated in preventive measures. The effect of these cannot be better illustrated

than in the notified cases of lead poisoning which, in 1900 were 1,058 and, in 1913, 535 – a reduction of approximately 50 per cent.[96]

In 1908, a Departmental Committee was appointed to inquire into the dangers attendant in the use of lead and the danger or injury to health arising from dust and other causes in the manufacture of earthenware and china. This Committee reported in 1910 that the time was not yet ready for a prohibition on the use of raw lead but occupiers should be encouraged to dispense with glazes containing more than 5 per cent soluble lead. The Committee also emphasised the danger of lung disease as a consequence of breathing air laden with dust in the potteries. The following year, another Departmental Committee reported on humidity and ventilation in cotton weaving sheds. During 1910,[80] both Medical Inspectors were fully involved in a number of problems. Dr Collis made inquiries into conditions in tin plate works, the physiological effects of a warm humid atmosphere on weavers and dermatitis in the manufacture of rolled tobacco. The Annual Report by the District Inspector for Derby[80] gives the following account of the problem of dust in an iron foundry:

> The condition in large pipe foundries is a matter of anxiety as regards dust, in these they work at high speed day and night in a veritable sandstorm. Long iron pipes are cast vertically in a circular pit from 6 to 50 at a time. In a pit working 24 hours, four consecutive casts would be made, several under the same roof. The total of work and amount of dust is both vast and almost continuous. The dust is given off at the loosening of each mould, removal of the core and uplifting of the casting as each hot pipe is forcibly removed, scaled and left aside. It is irritating and blackens the face . . . some wear respirators.

A more scientific approach to the many new problems facing the Inspectorate was evident by this time. A Principal Chemist of a Government Laboratory carried out an analysis of dangerous materials in response to the White Phosphorus Matches Prohibition Act 1908, which came into force in 1910. 325 samples were submitted for analysis, thirty of which were samples of match paste taken from lucifer match factories. Samples of pottery glazes were analysed where cases of plumbism occurred or where restrictions on the use of lead applied. Air sampling was carried out and it was reported that 1,993 samples had been carried out in bake houses, dressmaking and other workshops.

Dr Legge, during the First World War, inquired into the illness of toxic jaundice caused by tetrachlorethane, trinitrotoluene and arseniuretted hydrogen and the illness of aplastic anaemia caused by trinitrotoluol and benzol, which were being used in the manufacture of materials of war. Other investigations were carried out into the prevention of poisoning caused by the dope used for the varnishing of aeroplane wings.[82] The Police, Factories (Miscellaneous Provisions) Act, 1916 had an important clause, which changed the long-standing duty of occupiers to report all accidents to the Surgeon. This was no longer required and the latter was only called upon to investigate cases referred to him by the District Inspector. At the same time, Mr W.S. Smith, who was HM Inspector of Dangerous Trades, was assisting the Reserved Occupations Committee on the prevention of accidents and dust explosions in munitions factories and securing adequate ventilation in aircraft works. To assist him in this work, Mr Price, who was subsequently to become one of the first Engineering Inspectors was brought to London. Mr Smith continued this work with Dr Collis and in 1917 they presented joint reports on the effects of dust inhalation on workers making ganister bricks. As a result of this work, draft regulations were produced for the industry. Dr Collis's work on the effects of ganister dust led to the Refactories Industries Compensation Scheme. The Annual Report [83] acknowledged the important work done by Mr Smith on the ventilation of aircraft factories to minimise the effects of fumes from doping rooms. It is also recorded that he received additional assistance from another Inspector, Mr Hird, who also became an Engineering Inspector.

An Internal Committee was set up to look at staffing levels and in 1920 it proposed that the additions to the technical staff should include two Medical Inspectors, one of whom should be a lady. Dr Middleton joined the Department in 1921, bringing specialist experience on the occurrence of silicosis in industry. Dr Middleton was the first to carry out a systematic clinical survey in the grinding industry. He demonstrated the incidence of silicosis among these workers and established the criteria for further inquiries into other industries where workers were exposed to silica bearing dust, chiefly in the sandstone and pottery industries. These inquiries were instrumental in the setting up of schemes for compensation and periodic medical examination by the Silicosis and Asbestosis Medical Board. Later inquiries by another Medical Inspector, Dr Merewether, were to establish the relationship between asbestos and fibrosis of the lung in workers involved in the manufacture of asbestos products. This is covered in more detail in Chapter 11.

Dr T. Legge

In 1927, Sir Thomas Legge resigned his post as Senior Medical Inspector, after twenty-eight years service, on learning that the Government had failed to pass a Bill to prohibit the use of white lead. He had voted for the Lead Paint Convention which prohibited its use and which was unanimously passed by the ILO in Geneva in 1921. Dr Bridge was appointed as the next Senior Medical Inspector. Dr Bridge joined the Department in 1914 and retired in 1942, to be succeeded by Dr E.R.A. Merewether who was a member of the staff since 1927.

Exposure to hazardous fumes and gases

The years following the First World War saw a massive increase in the production of iron and steel and the development of new processes involving hazardous chemicals. These developments led to serious toxic and long term ill health problems for the workers who came into contact with the fumes and gasses arising from the manufacturing processes. Carbon monoxide was a particularly dangerous gas in steel works and power plants. In the early years there were many multiple fatalities caused by exposure to blast furnace gases. Producer gas from the charging and stoking of generating plant and accidental leakage into unventilated areas was another source of carbon monoxide poisoning.

Hydrogen sulphide was another dangerous gas that was often found when workers cleaned tar stills or tanks. Chlorine became very popular after the war because it was easy to transport in gas cylinders for use in the bleaching of flour. The use of this gas reinforced the drive towards the provision of suitable respirators and proper training in their use. Ammonia was another gas that became popular as a refrigerant. The high pressures involved in ammonia refrigeration plant made it dangerous when leakage occurred. Table 10.2 provides a brief summary for the period between 1913 and 1958 of some of the notified cases involving these gases.

New industrial diseases

In 1924, the Annual Report[90] refers to the disquieting findings of Sir Thomas Legge on the association of the disease epithelioma with the occupation of mule minders in the cotton trade. A Departmental Committee had been appointed to inquire into the number of cases reported and the very high death rate. It was feared that many more workers had already been affected, which would make it difficult to effect an immediate reduction in the number of reported cases. The same Annual Report refers to the work of Dr Middleton, who, at the request of a secretary of a trade union, had made inquiries into the physical effects of pneumatic tools. Initially he was unable to find a definite injury to health, although able to confirm what was already known, that an effect is produced on the fingers of certain workers – the so-called white finger illness. Dr Bridge and Dr Middleton continued their research into white finger illness by examining four masons who had been using pneumatic tools. Provisionally, they were able to conclude that some disability was produced when two features acted together; firstly the pressure of the tool impeding blood circulation to the fingers, especially the left hand, and secondly the action of cold,

Year	Annual Total		Carbon Monoxide		Hydrogen Sulphide		Chlorine		Nitrous Fumes		Ammonia		Benzol Napha etc.	
	Total	Fatal	Total	Fatal	Total	Fatal	Total	Fatal	Total	Fatal	Total	Fatal	Total	Fatal
1913	-	-	59	7	8	1	1		0	0	3	0	6	2
1914	-	-	62	9	22	3	2	0	9	2	4	1	4	2
1917	-	-	99	18	11	4	3		62	5	4	1	4	2
1918	-	-	54	13	7	1	4	0	27	7	6	1	7	4
1919	-	-	85	12	3	0	9	0	5	2	8	0	9	3
1920	-	-	56	9	13	4	8	0	9	3	0	0	12	1
1921	115	19	77	14	3	0	3	0	0	0	9	1	10	0
1922	-	-	111	14	12	3	11	0	8	0	8	1	25	2
1923	280	15	134	7	8	0	16	0	7	0	5	1	55	3
1924	-	-	107	10	11	4	20	0	10	1	1	0	26	0
1925	200	19	118	10	4	0	12	0	10	2	5	0	3	1
1926	-	-	101	6	3	0	13	0	5	1	5	1	4	1
1927	-	-	88	4	9	0	14	0	7	0	5	0	7	2
1928	167	19	81	9	9	3	17	0	6	1	12	1	7	2
1929	-	-	113	10	6	0	14	0	11	2	18	0	7	0
1930	133	21	26	5	5	0	5	0	5	0	2	0	0	0
1933	149	14	80	9	3	2	13	0	0	0	3	0	2	0
1936	153	12	86	8	5	0	7	0	6	0	3	0	6	1
1937	196	20	92	15	3	1	19	0	7	1	8	1	6	2
1938	190	27	98	14	10	7	17	0	14	1	4	0	4	0
1939	184	11	84	5	6	2	29	0	9	0	1	0	4	0
1940	583	31	102	20	9	3	39	0	236	2	4	0	6	1
1941	782	41	258	24	17	1	16	0	217	1	13	2	18	5
1942	776	25	249	11	9	3	52	1	220	2	6	0	12	5
1943	695	27	231	17	12	0	28	0	135	4	9	0	3	2
1944	450	25	209	21	7	0	59	0	55	1	8	2	5	0
1945	427	27	218	18	6	0	47	0	29	0	12	1	7	1
1946	243	13	117	11	4	0	30	0	13	0	11	1	5	1
1947	244	22	131	8	14	7	22	0	10	1	11	0	2	0
1948	229	14	120	9	10	1	26	1	8	0	4	0	2	0
1949	218	24	100	18	11	2	26	0	13	0	8	0	1	0
1950	242	30	113	22	2	0	25	0	5	0	10	2	4	1
1951	228	17	105	9	12	3	21	0	3	0	8	1	1	0
1952	265	17	108	9	9	1	27	1	10	0	9	0	3	1
1953	254	23	121	19	4	0	36	0	11	1	11	1	8	0
1954	237	21	113	14	13	2	25	0	14	1	9	1	6	2
1955	255	11	107	7	5	0	24	0	15	0	21	1	11	3
1956	236	18	89	11	15	1	21	0	6	1	12	1	8	2
1957	232	22	76	10	6	3	30	0	4	0	7	1	3	1
1958	218	12	79	9	4	0	32	0	4	0	4	0	0	0

TABLE 10.2: Notifiable cases of accidents caused by gas or fumes from 1913 to 1958

caused by the escape of air acting on parts to which the blood supply is impaired. They could not find sufficient evidence to show that serious permanent injury to the nervous system or to general health was likely to follow. Compressed air illness, or caisson disease, was investigated following three cases of workmen being affected while working in a narrow bore tunnel under the River Thames at pressures of 30 to 35psi. In each case the men had decompressed too quickly. Assistance was obtained from the Admiralty and the Diving School at Portsmouth, which resulted in a recommendation that workers should be medically examined and carefully selected before taking up this work.

In the same year, Dr Middleton and Mr Macklin were responsible for draft regulations for the Grinding of Cutlery and Edge Tools and the Grinding of Metals (Miscellaneous Industries). The Compensation Act 1923 abolished the dual reporting standard of one-day absence in the case of machinery accidents and seven days otherwise. Only a three-day inability to earn full pay, without differentiation as to cause, now became the basis for reporting cases of injury and ill health to the Factory Department.

Two new industrial illnesses first reported from the USA, attracted the attention of the Medical Department in 1925.[91] The first report described necrosis of the jaw and aplastic anaemia, which had resulted in five fatalities. This affected women employees using luminous paint that contained radium salt. Dr Bridge established that in this country, the paint used in the process contained zinc sulphide with a small proportion of radium bromide added to improve the phosphorescence of the zinc sulphide. Examination of the women working with these materials did not reveal symptoms comparable to the USA. Blood samples from seven women working with these radioactive substances for periods from 1 to 10 years did, in three cases, show some change that was attributed to radium salt. It was concluded that the illness from luminous paint probably happened because the material had been swallowed through putting the paint-brush in the mouth.

The second report concerned the manufacture of tetraethyl lead and its use as an additive to petrol. There was general apprehension about the possibility of lead poisoning affecting members of the public exposed to exhaust fumes. In the USA, 149 cases including 11 deaths had occurred in the course of manufacturing this organic lead compound. The USA research indicated that the danger was limited to the process of blending and ethylising the petrol during manufacture, when the compound could be absorbed through the skin. It was concluded that there was no good grounds for prohibiting the use of ethyl gasoline provided that its distribution and use were controlled by proper regulations.[54] Concern about the use of ethyl petrol in 1925 resulted in the Ministry of Health setting up a Committee to inquire into its possible dangers to health. Dr Bridge was a member of that Committee, which in 1928 published an interim report. By this time, Dr Bridge was the Senior Medical Inspector and he had this to say in his Annual Report:

> Increasing attention to the effects of industry on health of the worker . . . not only industries where systematic poisoning is produced by use of recognised toxic substances, or a localised disease caused by inhalation of a dust not in itself toxic, but also physiological and psychological effects from fatigue and monotony. The simplification of industry reduces physical energy but introduces new factors . . . Chemical substances hitherto unknown

outside research laboratories are being used in industry with little or nothing known about their effects . . . there is a need for careful observation.[92]

By 1930, the workload of the Medical Branch was eased by the addition of two new Medical Inspectors. A number of pamphlets and publications were issued during the year, which were important contributions to the list of Departmental guidance on industrial health problems:

> Asbestos Dust, Report on Effects of Asbestos Dust on the Lungs and Dust Suppression in the Asbestos Industry. by Dr E. Merewether & Mr C.W. Price.
> Chrome Plating. Memorandum on Chrome Plating and Anodic Oxidation.
> Cellulose Solutions. Memorandum on the Manufacture, Use and Storage.

In order to give effect to the recommendations of the Silicosis (Medical Arrangements) Committee, the Workmen's Compensation (Silicosis and Asbestosis) Act, 1930 was passed, giving powers to the Secretary of State to arrange schemes for the medical examination at regular intervals for purposes of compensation of workers in certain industries and processes giving rise to the dust of silica and asbestos. This came about because at last, the recognition of fibrosis of the lung due to asbestos dust as a disease of industrial origin could no longer be disputed. It was hoped that steps taken in regard to prevention and compensation would be followed in other countries. As will be seen in the next chapter, there was a fatal flaw in the steps taken to prevent this disease, which is only now becoming apparent.

The increasing complexity of industrial health problems meant that the Factory Department had to increasingly seek expert assistance from outside bodies, particularly in the field of research. By 1935,[100] the Department of Scientific and Industrial Research were involved in a number of projects such as:

> Detection and estimation of toxic gases and standardisation of canister respirators at their Chemical Defence Research Department.
> Standardisation of respirators for protection against dust at CDRD.
> The toxic effects of volatile substances, under the Medical Research Council.
> Problems with the inhalation of dust, under a committee on industrial pulmonary diseases.

By 1937[101] the Senior Medical Inspector was able to report that the investigations had continued on methods of detection and estimation of small quantities of certain toxic gases present in air. Leaflets were published on appropriate methods for the detection of hydrogen sulphide, hydrogen cyanide and others relating to aniline vapour, benzene vapour, nitrous fumes and sulphur dioxide were about to be published. The Chemical Defence Research Department at Porton Down completed the design of a very efficient dust respirator that was to be made available to industry.

The 1937 Act contained new health provisions which prompted Dr Bridge in his report[102] to question to what extent the new provisions would improve the health of industrial workers and to what extent the positive requirements concerning cleanliness, over-crowding and ventilation would improve health and reduce the incidence of infections such as the common cold. However his greatest concern was for the effect upon health from the inhalation of dust. Dr Bridge's general concern for emerging health problems was repeated in the 1940 Annual Report,[104] which said:

> The production and use of various other chemical bodies, sometimes new and sometimes in great quantity, add to the specific diseases, which have occurred but cannot be referred to in detail. Every aspect of the cause and prevention of such cases is, however, most carefully considered. There is a marked increase in the number of voluntary reported cases of dermatitis (4744 compared with 2952). The most significant increase is in chemical manufacture (1298 compares with 477) due to inexperienced workers. The increasing use of radioactive paint has required special consideration, precautions are effective but close supervision is being continued.

The demands of the war and the sensitive nature of some of the industrial materials and processes meant that the Medical Branch were involved in a critical activity of balancing the health needs of the country's workers against the exceptional demands made upon them to meet the needs of war. Dr Bridge, who joined the Department at the beginning of the First World War in 1914, retired on 31 December 1942. Dr E.R.A. Merewether, who had joined in 1927, was appointed as the new Senior Medical Inspector of the Medical Branch, which had now grown to thirteen in number.

CHAPTER 11

THE LEGACY OF ASBESTOS

The cost of exposure to asbestos

Industrial and environmental exposure to asbestos dust during most of the last century left a terrible legacy of chronic illness, premature death and a financial cost upon society. The insurance industry was burdened by claims for personal damages caused by past exposure to this hazardous material. Claims made in 1985 for asbestosis, an occupational disease, reached 500 a month. By 1988 the claims had increased to 2,000 each month. In 1993 the London insurance market paid out £2 billion in settlement with many more claims still in the pipeline. The global financial burden facing the companies covering this type of risk is well documented and the liability costs were calculated in terms of personal hardship, bankruptcy, and claims of negligence. It was estimated in 1994 that the insurance industry would pay out £10 billion over a period of twenty years.[55] Over-exposure to asbestos is a fate the insurance industry share with many thousands of workers and ordinary people who came into contact with this deadly material during the course of their daily lives.

Asbestos dust is well known as a cause of asbestosis and lung cancer, and as a primary agent in the development of the modern industrial disease called mesothelioma. The risks were apparent from the established epidemiological records of asbestos related illness and the long-standing concerns about the industrial use of this material. Sadly, the current level of asbestos-related disease could have been reduced if the hazard had been acted upon at the appropriate time. It has been reported[56] that asbestos will outstrip motor accidents as a cause of premature death, with over 3,000 already dying each year. Statistics predict the death toll would rise to 10,000 by the year 2020 because of exposure that took place in the 1960s and 1970s. It is known that mesothelioma killed 200 people in 1971. This rose to 1,017 in 1992 and continues to rise. This incidence of fatalities far exceeds the current annual death rate for all industrial accidents in the United Kingdom.

Why has asbestos exacted such a high price on the health of so many working generations? The answer can be obtained from the personal experiences of those who came into contact with asbestos in the past and from other records of asbestos as a source of occupational illness in the twentieth century. My personal experience of asbestos began in Glasgow in 1949 at the start of my working life. This was a time of high industrial activity as the country recovered from the ravages of the Second

World War. At that time Glasgow was a great world centre for shipbuilding. Shipyards stretched along the length of the river Clyde from the centre of Glasgow to its lower reaches at Greenock on the south bank and Dumbarton on the north bank. There was no difficulty for school leavers to obtain employment and an apprenticeship in the many trades that serviced the heavy industries around the city.

I started my first job as an apprentice draughtsman in a small engineering company on Clydeside. The company was typical of the many sub-contractors providing services to the shipyards and other engineering works. It was ideally situated in the heart of a shipbuilding and residential district where great shipbuilders like Barclay Curle, Fairfields, Yarrows and John Brown's were within easy reach. The company was part of the Bestobell group and was a specialist contractor that carried out all types of insulation work. Insulation was a vital part of the engineering scene, steam boilers, pipework and machinery had to be clad with thermal insulation to improve the efficiency of the plant. Factory and office walls and roofs had to be insulated to conserve energy, which was still in scarce supply at that time. Ocean-going liners and cargo ships had to be equipped with insulated and refrigerated cargo spaces to facilitate the transport of food and other vital commodities.

The first eight months of my employment was spent in the workshop as a store boy. The factory was a conglomerate of different activities and trades. A metal fabrication shop made many different component parts for the ships. Blacksmiths

Raw asbestos stacked at Dalmuir works of Turners Asbestos Cement Co. Ltd

worked alongside welders, fitters and machine operators; cutting, drilling and assembling parts for various contracts. The woodworking shop employed skilled shipwrights and joiners making quality wooden fittings for the ships. Prefabricated pipework and gas cylinders in racks were assembled in a plumbers' shop for engine room fire extinguishing systems. The whole factory was a concentration of activity of skilled tradesmen with a lifetime of experience in the proud business of building ships. The works were located alongside and surrounded by tenement homes that were typical of the city's residential areas, and of course these homes were in close proximity to any asbestos dust emanating from the activities of the premises.

A general store was located at the far corner of the yard. Many materials used by the tradesmen were held there until needed. The largest area of the store was set aside for bulk quantities of insulation materials. Cork in granulated or slab form, bags of vermiculite, pre-formed pipe sections of magnesia and sacks of asbestos fibre were stacked there. Crocidolite asbestos (blue asbestos) was a major part of the company's business, used under the trade name of Sprayed Limpet Asbestos. The asbestos was unpacked and stored in bays, until it was required, mainly for the application of fire protection on ships bulkheads. The material was soft, slightly warm to the touch and offered no clue to its lethal legacy. Like many others at that time I handled the material, became covered in its dust and watched it drift freely from the store to deposit around the factory and beyond.

Insulation mattresses used for lagging pipework were made by hand on the premises. The mattresses were made from two layers of asbestos cloth stitched together with asbestos twine and filled with loose asbestos fibre. At no time was I or anyone else made aware by my supervisors or outside agencies of the potential danger from the

Workmen fixing insulation slabs to a structure

dust, nor was I instructed in the precautions imposed upon its use by any health and safety regulations. If someone knew about the risks they never said, and if the local Factory Inspector visited the premises during that period, so far as I was aware, no action was placed upon the management to improve the working conditions in the works, the local environment, or at the other sites where the insulation was used.

A technical sales book on heat insulation[57] published in the 1950s by a specialist contractor, provides an example of the detailed information available on insulation technology and materials, including the different types of asbestos. Yet there is no mention in the booklet of any health hazard from these materials or of any precautions to be taken to reduce the risks from exposure. This is what it says about asbestos:

Asbestos

Asbestos is a collective name for naturally occurring minerals of a fibrous nature, the various types having a wide range of strength, flexibility and quality of fibre.

Three types are widely used in the preparation of thermal insulation materials, namely: Amosite, Chrysotile and Crocidolite, the suitability of the fibre being governed by its staple length, tensile strength and resistance to heat and chemical action.

Amosite is a long staple asbestos found chiefly in Africa and has a relatively high iron and water of crystallisation content. Dehydration occurs between 1,000°F and 2,000°F, and some discoloration is evident due to the oxidation of ferrous or ferric iron. Dehydration is accompanied by embrittlement of the fibres, and fusion takes place at 3,030°F.

Chrysotile is a soft white fibre found mainly in Canada and forms the bulk of the world production of both long and short fibres. Dehydration occurs between 1,000° and 1,400°F and fusion at 2,850°F. Since the iron content is low, dehydration proceeds without change of colour, the resulting oxides being mainly magnesium, which is white. Embrittlement of the fibres is apparent during the dehydration period.

Crocidolite or Blue Abestos, is similar in composition to Amosite although dehydration occurs between 800° and 900°F and fusion at 2,080°F.

The asbestos ore occurs in three types of deposits: Cross fibres, slip fibre, and mass fibre. The deposits of Amosite, Chrysotile and Crocidolite are the cross fibre type where the fibres occur in seams with the fibres extending across the seam from wall to wall. The width of seam varies from a mere streak to a maximum of 6ins. In slip fibre deposits the fibres lie parallel to the seam walls and are broken where rock movements have created shear zones.

In mass fibre deposits the entire rock is composed of bundles of fibre.

Following excavation the ore is crushed and the asbestos fibres cleaned, teased and graded according to fibre length.

The medium length fibres of Amosite and Crocidolite are used in the manufacture of preformed pipe sections and slabs.

The short length fibres including Chrysotile are felted into paper and millboard or used as a binder for diatomaceous silica, 85 per cent Magnesia, insulating and hard setting cements, and many of the preformed and moulded insulation materials.

Sprayed blue asbestos was a popular way of fireproofing ships. Large surface areas could be covered quickly and cheaply. The same technical book extolled the merits of sprayed asbestos, and as above, there was no mention of any health hazard from this most hazardous type of asbestos or of any precautions to be taken.

When mixed with suitable binders asbestos can be applied to building surfaces by a spraying process. Essentially an asbestos spraying plant consists of a carding and blowing machine from which the fibre and air are delivered through a smooth bore rubber hose to a spray gun, which introduces a fine spray of water to the fibre and discharges the damp mixture on to the surface to be insulated. The asbestos layer is rolled and tamped to the desired thickness and allowed to dry naturally. The resulting layer is light in weight and has an unbroken surface, which is advantageous when an anti-condensation treatment is desired.

Anyone with experience of the hectic pace of ship construction, particularly when being fitted out, will appreciate the problems presented by sprayed asbestos in the many confined spaces. Tradesmen competed for access to erect steelwork, find space for ventilation ducts, electric cableways and pipework. Steel bulkheads were no sooner erected than they had to be cut or modified to obtain access. Although the spraying would be done at night, there was no way that subsequent exposure could be prevented, and in truth no such effort was ever made in those early days. Workers came in and out of these confined spaces that were covered on all sides by slowly drying blue asbestos. It might be concluded from these recollections that the 1950s and '60s was a period of ignorance of the risk from exposure to asbestos, and it might be concluded therefore that the present level of lung diseases was unavoidable − this is far from the truth. The story of asbestos, its value as an industrial material *vis-à-vis* its toxicity and the actions of vested interests and government agencies provides a lesson and a warning that should not be ignored. Some of today's chemicals, industrial materials and environmental agents, such as microwave radiation, with an as yet unproved toxicity, may well turn out to be tomorrow's equivalent of asbestos.

An early awareness of the risk

Occupational lung disease is as old as the Industrial Revolution. As far back as 1861, Sir John Samon, the Medical Officer of the Privy Council, inquired into the death rates from lung diseases in industrial occupations. He concluded: 'the canker of industrial diseases gnaws at the very root of our national strength.' As a result, the Factory Act of 1864 was amended to require factories to be ventilated so as to render harmless, so far as practicable, any gas, dust or other impurities that might be injurious to health. The first recorded fatal case of fibrosis of the lung in an asbestos worker occurred in 1900 and was described by Dr H. Montague Murray in evidence before the Departmental Committee on Compensation for Industrial Diseases in 1906. The Medical Inspector's report in the 1910 Annual Report,[80] which was addressed to Winston Spencer Churchill the Secretary of State for the Home Department, makes the following observations on asbestos:

Following up information from the Register General it was found that 5 deaths of persons suffering from phthisis had occurred in 5 years among a staff of 40 workers employed at a factory where asbestos is woven. The process which appears most dangerous is the production of asbestos mattresses which are composed of bags of woven asbestos filled with short asbestos fibres, they are placed on a table and beaten out flat by a man with a wooden flail which causes dust to rise. Women who sew the mattresses into sections with asbestos threads worked close to the men who beat the mattresses and of necessity inhale the dust. The reorganisation of this process with the application of localised exhaust draught was called for and an annual medical examination by the Certifying Surgeon has been instituted, in the hope of detecting and removing from exposure to dust those showing early signs of respiratory disease. Weaving asbestos has become an important industry during the last 15 years.

In 1924, Dr W.E. Cooke announced the discovery of 'curious bodies' found on microscopic examination of lungs. Post mortem examination showed fibrosis of the lungs and change due to pulmonary tuberculosis. In February 1928 Dr McGregor, the Medical Officer of Health for Glasgow, drew attention to an asbestos worker receiving treatment for pulmonary fibrosis, which might be connected with employment. An enquiry was undertaken to prove or disprove the existence of a risk by Dr Merewether, HM Medical Inspector in the Factory Department. He undertook a survey of carding, spinning and weaving in the textile industry. Intensive scientific and medical evidence was now being obtained to prove without doubt the link between asbestos and fibrosis of the lung. In 1928, Dr F.W. Simson reported a fatal case of fibrosis of the lungs in a worker in an asbestos mill in Southern Rhodesia. Professor Stewart of the University of Leeds also showed that the 'curious bodies' could be obtained during life by puncture of the lung in asbestos workers and in the sputum and after death in the lung juices. These developments led to the following observation in the 1928 report.[92]

> With justice it may be asked why is it that this disease has only recently attracted notice and become a problem in this country. A number of factors have contributed to obscure the existence of such a disease amongst workers in asbestos, of which the most important seem to be the comparatively small number of workers engaged in essentially dusty processes, the insidious nature and generally slow progress of the disease, and the difficulty of accurate diagnosis. Furthermore, the instructions by Inspectors to press for enclosure of, or the application of exhaust ventilation to dusty processes in the industry, must have checked the earlier incidence by materially reducing the concentration of dust. It is noteworthy that the majority of fatal cases, which so far have come to light during the present inquiry have occurred in workers in a process in which the application of efficient exhaust ventilation presents pronounced technical difficulties.

The joint report by Dr Merewether and Mr Price in 1929[93] provided the first conclusive scientific evidence of the link between fibrosis of the lung and exposure to asbestos and the disease that became known as asbestosis. Their epidemiological survey of 374 workers discovered 105 workers with diffuse fibrosis of the lung caused by the inhalation of dust, including asbestos. Radiographic examination of 133 cases confirmed fifty-two had fibrosis of which twenty-two showed signs of early change in their lungs due to asbestos. Ninety-five cases of asbestos fibrosis were examined in respect of their age, length of employment and work process. This showed that the risk varied with length of employment. There was no definite case of fibrosis due to asbestos in workers with less than five years exposure. For longer periods of exposure the incidence rates ranged from 25.5 per cent in five to nine years to 80.9 per cent in twenty years and over. Merewether and Price concluded that the disease could be fully developed after seven to nine years and be the cause of death after thirteen years' exposure.

So by 1929, the existence of asbestosis was fully established and the serious toxic nature of asbestos dust over a relatively short working period of time proven. Dr Merewether summarised his conclusions in the following graphic description of the disease:

> The outcome of this inquiry is to establish without doubt the occurrence of a definite type of fibrosis, differing in character from that produced by free silica dust, and giving a radiological picture distinct from that of true silicosis and that the damage to the lung is much greater than it appears to be. The asbestos fibrosis is much more diffuse; it spins its fine web, as it were, criss-cross through the lung, enveloping and eventually strangling the ultimate lobular structure, rather than depositing itself in numerous more or less isolated foci as in silicosis. Thus in less advanced stages the radiological picture does not impress the eye, unconsciously viewing it from the standpoint of silicosis. It is true that the picture may show soft and fairly course nodulation, but it is never so impressive as in the silicotic film. Paradoxically the distinctive feature is its uniformity; the unobtrusive but diffuse impairment of the percussion note; the homogeneous stripping of the skiagram, are all fragments of an entity, unmistakable when assembled, but enigmatic when divorced.

The Factory Inspectorate was quick to move on these findings.[94] Textile manufacturers were invited to a meeting at the Home Office where agreement was reached on the need to suppress dust. A committee was set up under the chairmanship of Mr Leonard Ward, the Senior Engineering Inspector of the Engineering Branch. The other members of the committee were:

Mr J. Gow of Cape Asbestos Co. Ltd.
Mr P.G. Kenyon of Turner Brothers Asbestos Ltd.
Mr W.C. Fenton of British Belting Asbestos Ltd.
Mr C.W. Price, HM Engineering Inspector.

The principal safeguard was considered to be improved ventilation in the factories where the asbestos was processed. The manufacturers did, however, register concern that not all processes were dangerous. Their case was based upon the assumption that there was a critical lower limit of dust concentration below which workers could be employed without injury to health. They suggested that conditions in flying spinning without exhaust ventilation might be regarded as a 'dust datum subject of course to alteration in the light of further medical experience'. The recommendations of this committee were confirmed at a second conference of manufacturers in March 1931. These were submitted to the General Council of the Trades Union Congress who made further suggestions to be taken into account in drafting a code of regulations under the 1901 Act. The Chief Inspector of Factories, Mr Gerald Bellhouse, made the following observation in the 1930 Annual Report:

> In view of Dr Merewether's findings, the necessity of dust suppression became obvious and urgent. He and Mr Price were detailed to study this. Results were published in a joint report with a view to expedite a Code of Regulations, which would be acceptable to

industry ... The agreed recommendations should be effective in reducing the incidence rate of pulmonary fibrosis. The whole hearted desire on this section of industry to grapple with the occupational risk is evidenced by the clarity with which he has tackled the problems and the way in which individual firms have generally pooled the results of much experimental work.

The engineering solution of ventilation and dust suppression being developed at that time related only to manufacturing processes. There was less recognition of a wider problem when asbestos was put to use after manufacture. The thorough scientific analysis of the disease was now seen as being important. The Senior Medical Inspector in his report[94] noted:

If industrial medicine is to progress as a scientific branch of medicine, the study of conditions found in industry must go hand in hand with laboratory work. Correlation of the results of practical medical inspection with laboratory investigations will enable a more correct assessment of the responsibility of industry for clinical conditions to be made. A syndrome of symptoms accompanied by facts established in the laboratory are essential to establish definite industrial risk.

Medical investigation commenced in 1930 with the records of twenty fatal cases of asbestosis. This confirmed the grave risk that accompanied continual exposure to heavy concentrations of dust. Of the twenty cases examined, six were mattress makers, six were carders or cloth weavers and in the remainder, evidence pointed to dust arising from nearby dusty processes. There was evidence that removal from the process delayed the appearance of a disabling fibrosis for many years, but it seemed clear that exposure to heavy concentrations of asbestos dust for a short period of years resulted in the appearance sooner or later of a disabling fibrosis. It appeared probable that the product of two factors − concentration of dust in the air breathed and length of exposure − was within limits a constant. The average age at death of the twenty cases was 38.9 years and average length of employment 14.9 years.

On the basis of these findings the Inspectorate carried out a lengthy study of asbestos related deaths from 1931 to 1956, which is summarized in Table 11.1. This covered 271 deaths from asbestosis and a further ninety deaths from asbestosis complicated by tuberculosis. In 1934[98] the data showed that in the case of asbestosis, the average age at death was 41 and the average duration of employment was 12.9 years with a span of employment between 1.5 and 27 years. By 1956[114] the average age at death had increased to 50.5 years and the average duration of employment had increased to 16.7 years. Significantly, the shortest period of employment was by then only six months, a figure that remained unchanged in the statistics from 1943 onwards. The Inspectorate concluded from these statistics that the Asbestos Industries Regulations of 1931, with its requirement for exhaust ventilation in factories, had been beneficial in reducing the risks of exposure to asbestos dust.

The increasing use of asbestos in processes other than its manufacture was noted as far back as 1932.[96] An investigation was made by the Inspectorate into the atmospheric dust contained in a new process of spraying dust fibre together with an

Year	Asbestosis					Asbestosis with tuberculoses				
	No of Cases	Ave Age	Exposure – years			No of Cases	Ave Age	Exposure – years		
			Shortest	Longest	Average			Shortest	Longest	Average
1931	24	40.6	4.4	28.0	15.1	11	45.7	2.6	18.0	13.5
1932	27	40.8			15.2	15	41.6			11.7
1933	35	41.0	3.5	27.0	13.4	18	38.3	2.3	18.0	9.6
1934	41	41.0	1.5	27.0	12.9	26	38.1	0.8	29.0	9.0
1935	52	41.9	1.5	27.0	12.4	30	37.1	0.8	29.0	9.5
1936	59	42.8	1.5	36.0	13.3	34	37.2	0.8	29.0	9.6
1937	68	43.5	1.5	36.0	13.4	38	38.2	0.8	29.0	9.4
1943	118	46.4	0.5	48.0	15.1	62	39.0	0.8	29.0	10.4
1944	125	46.6	0.5	48.0	15.1	65	38.8	0.8	29.0	10.4
1945	134	46.9	0.5	48.0	14.9	67	39.0	0.8	29.0	10.2
1947	160	47.5	0.5	48.0	16.9	72	39.0	0.8	29.0	10.4
1948	171	48.1	0.5	48.0	15.6	76	39.0	0.8	29.0	10.3
1949	186	48.2	0.5	48.0	15.6	78	39.2	0.8	29.0	10.5
1950	197	48.6	0.5	48.0	15.6	80	39.1	0.8	29.0	10.6
1951	211	49.1	0.5	48.0	16.4	82	39.4	0.8	33.0	10.8
1953	230	49.5	0.5	48.0	16.6	87	40.2	0.8	33.0	11.4
1954	237	49.6	0.5	48.0	16.5	89	40.5	0.8	33.0	11.4
1955	257	50.2	0.5	48.0	16.6	89	40.5	0.8	33.0	11.4
1956	271	50.5	0.5	48.0	16.7	90	40.7	0.8	33.0	11.5

Table 11.1: Asbestosis mortality study – 1931 to 1956

adhesive on to the walls of a railway tunnel. Results showed the protective power of uni-directional ventilation in confined spaces. This conclusion seemed to overlook the consequences of the ventilation system directing the dust into the environment and towards unsuspecting persons. The report also noted that the process had been extended to the coating of the inside of the bodies of motor vehicles.

It was realised very quickly that there was inevitability about the consequences of the disease. It was believed that ignorance in the past had allowed the disease to take hold of those unfortunate enough to be exposed to asbestos dust. This inevitability was rather clinically expressed in the 1933 report,[97] setting an often-repeated theme for future commentaries on the development of the illness:

> Disablement through its agency, are due to long past exposure to unfavourable conditions. It follows therefore that in our present state of knowledge, prevention must inevitably be preceded by the sacrifice of some health and even of life on the part of individual workers and the main aim must be to reduce the number of such victims to a minimum.

Throughout the 1930s dust and its effect upon the health of workers continued to be one of the greatest problems facing industry. The Senior Medical Inspector acknowledged this in his 1938 report:[102]

We are but on the threshold of knowledge of the effects on the lungs generally and I have referred in my reports from year to year to the enquiries made into cases of illness and death alleged to be due to inhalation of dust. Moreover dust thought today to be harmless may, following research, be viewed in another light tomorrow. It is not many years ago when the dust of asbestos was regarded as innocuous, today it is regarded as highly dangerous.

Asbestos and the link with cancer

Cancer of the lung was linked with silicosis and asbestosis in 1938. Among 943 deaths, cancer of the lung was found in twenty-three cases. In seventeen of these cases cancer was the primary cause of death. Among 347 deaths from silicosis the post mortem showed cancer of the lung in seventeen cases. From a total of 103 deaths from asbestosis, cancer of the lung was revealed in twelve cases. These findings set in motion a new line of medical survey. By 1966 the incidence of lung cancer in association with asbestosis deaths was 36.9 per cent for males and 17.5 per cent for females.[116] This link with cancer was set out in the Annual Report for 1955, which tabulated the cause of deaths of 365 male and female workers with the three manifestations of this disease, asbestosis, asbestosis with pulmonary tuberculosis and asbestosis with cancer of the lung. These statistics and the data contained in Table 11.2 can be seen as the seeds of a future disaster of much larger proportions.

Diagnosis	Male	Female	Total
Asbestosis	127	85	212
Asbestosis and Pulmonary Tuberculosis	47	41	88
Asbestos and Cancer of the Lung	48	17	65
Total	222	143	365

Table 11.2: Asbestosis deaths recorded – 1924 to 1955

The coming of the Second World War probably relaxed the occupational health concerns about asbestos. It was now a vital commodity needed for the war effort. Statistics kept during the war years showed a fairly constant trend in the number of deaths from silicosis and asbestos. By 1944 the Senior Medical Inspector was noting once again that records of cases of industrial gassing, industrial poisoning and lung disease, were giving cause for concern when compared with the general run of accidents. For each recorded case, he observed there were unknown numbers of people affected to a lesser extent by the same occupational hazard to their own detriment and to that of the State. He noted:

This is a chastening thought and a continual spur to our vigilance and our imagination. We must pay not less but more attention to the elimination of occupational disease. The Factory Department must always record the past in terms of lessons for the future; lessons recorded as much for ourselves as for the public.[108]

At the end of the war in 1945 the Inspectorate was able to take stock of the asbestos situation. The Annual Report[109] noted the development, during the war years, in the use of asbestos for heat and sound insulation aboard ships, with consequent increase in the number of workers exposed to risk of injury to health through asbestosis. They acknowledged that this now called for action. They stated that if the risk continued, they would consider new statutory requirements in amended Factories Act regulations. This intent was not carried out; instead they sent a letter to employer organisations and others calling attention to the matter and suggesting certain precautions. The Shipbuilding Employers Federation and the Trades Unions accepted the recommendations. The Inspectorate was now satisfied that adequate action had been taken to reduce the risk and improve working practices in these industries. Clearly their hopes were not to be realised and asbestos dust was set free to do its worst on a much larger exposed population.

A strange account is given in the 1945 report[109] of experimental work on asbestos, which was sponsored by one of the large asbestos manufacturers. The object of this research was to determine if aluminium mixed with asbestos had any retarding effects on the development of pathological lesions due to asbestos. One of the outcomes was the conclusion that aluminium modified the picture only slightly but the rats receiving aluminium plus asbestos appeared to live longer than those receiving asbestos only.

Despite the absence of positive enforcement action beyond the 1931 Asbestos Industries Regulations, the Inspectorate continued with its medical surveys and periodic warnings about the scale of the problem. In 1945[109] they reported:

> In respect of certain occupational diseases, e.g. silicosis there is a long lag between changes in environmental conditions and improvements in preventive measures and their reflection in incidence and mortality rates. Another aspect which often goes unappreciated and is not comforting is that the population exposed to a particular risk may be very small, only a few hundreds, and one or two deaths each year from a preventable disease represents an alarming incidence. Much thought is rightly given to the appalling number of deaths from road accidents, but a little calculation indicates that, taking into account the size of the populations at risk, the risk of death or disablement from certain occupational diseases in the appropriate populations is several times higher.

The use of asbestos spraying techniques again attracted the interest of the Inspectorate, sufficient to record the following in the 1949 Annual Report:[113]

> Portable asbestos spraying plants are used to an increasing extent by contractors on buildings and ships, etc., for insulation purposes. Asbestos is fed into the spraying machine by hand, and conveyed by means of a small spiked band to a small card cylinder, which takes the fibre from the travelling band and feeds it in a flocculent condition to a fan by which it is blown to a spray gun where it is ejected by means of compressed air. The spray gun

consists of a centre jet for asbestos surrounded by four jets of water, which converge on the asbestos spray at about one foot from the end of the gun. Also a fine spray of water is provided over the spike band conveyor in the actual machine. The control by means of a valve on the spray gun is so arranged that the water jet must be in operation before asbestos can be delivered to the gun. If the water supply fails the supply of asbestos is automatically stopped. In view of the risks to health, the firm from which the equipment is hired runs a school where employees of various insulation contractors are given a course of training in their proper use and manipulation, lasting about 14 days, and representatives attend these courses from overseas as well.

It is a matter of some interest that this report coincided with the personal experiences recorded earlier in this chapter. The earliest lessons of the 1930s with the need to provide proper ventilation and dust suppression seemed to have been forgotten now. The spraying technology was quite inadequate in this respect. The application of asbestos in this way was free from the strict controls that existed in the manufacturer's premises. The reality of the industrial situation was such that very few people knew there was a problem and those that supplied the material and equipment had there own reasons for keeping this so. It is most probably from this base that the present occupational health tragedy was allowed to grow.

The final chapter

The final chapter in the progress of asbestos-related diseases began in 1960[115] when evidence pointed to an association between asbestosis and a relatively rare tumour known as mesothelioma. Epidemiological evidence also showed that the incidence of mesothelioma was higher in non-occupationally exposed populations living in the vicinity of asbestos using factories, and also pointing to crocidolite (blue asbestos) as the particular variety of asbestos responsible. This new development resulted in the Factory Inspectorate appointing an Advisory Panel on Asbestos to advise the Senior Medical Inspector on existing medical knowledge on the hazards of asbestos. Work was started on revising the 1931 regulations. Mesothelioma had now been identified as the cause of 200 deaths over the previous fifteen years. Public interest reached a climax with the publication in 1965 of an epidemiological survey that indicated a significant excess of the disease in the vicinity of asbestos factories in London, and also in circumstances suggesting that asbestos contaminated clothing introduced into the home might be significant.

A medical register was set up in 1967 to record cases of mesothelioma. Between 1967 and 1969 a total of 378 cases had been registered from which it was observed that a brief exposure of one month could be associated with malignant mesothelioma half a century later, or death might take place in persons as young as twenty-eight years of age in association with high exposure to occupational asbestos. It is noteworthy how close these findings were to the earliest conclusions in the 1930s, which linked asbestosis deaths with the duration of employment.

In 1967 the Advisory Board published a memorandum that recommended further study of the mesothelioma problem. The objective was to determine the natural history of the disease by examining the relationship to asbestos exposure, the nature of the exposure and the development of asbestosis in an exposed population. Data prepared by the panel from 430 certified cases of asbestosis over a period of ten years showed that there was an increased mortality from the disease of about two to three times that of the general male population of equivalent ages. The Advisory Panel met twice in 1970 and agreed that a combination of a mortality study with a study of radiological changes would be a practicable method of pursuing the objectives. Evidence pointed to the significance of the smoking habits of asbestos workers. A start to the study was made in 1971 at HM Dockyards in Devonport where chest radiographs of 100 workers were examined. This showed that in the earlier radiographic films taken of each worker, ninety-five were normal and five abnormal, whereas of the more recent films, only thirty-five were normal and sixty-five abnormal.

Following the implementation of the new Asbestos Regulations in 1969, the Inspectorate increased its level of activity and visited more than 300 factories to make measurements. Sampling methods were used to count the asbestos fibres in the workers' breathing zone and a 'hygiene standard' was set for this exposure. Certain types of industry were singled out for special attention. It was acknowledged that although crocidolite was little used in factories now, exposure to the material would remain for many years in buildings where it had been used in the past. The construction industry was now seen as a source of high risk because of the widespread use of sprayed asbestos. Its use was still being condoned by the Inspectorate, which noted that the industry had now turned to a method of pre-damping prior to spray application which when properly carried out would reduce respirable dust a hundred-fold. This, they concluded, would enable members of the spray team to be fully protected by wearing simple approved dust respirators, and other tradesmen working outside would not be exposed to concentrations above the hygiene standard.

The worst situation now facing industry was identified as demolition work, where large quantities of asbestos needed to be removed. Crocidolite remained a problem in view of the severe hygiene standard (one-tenth of the concentration applied to other forms of asbestos). The solution was now proving very costly, with extensive precautions being necessary to comply with new Regulations, such as thorough soaking and removal of the asbestos from site in sealed bags. A high standard of personal respiratory protection was now a requirement.

Environmental studies and asbestos surveys were well under way by 1972.[117] The studies involved sampling the atmosphere over a period of four hours by personal sampler on one man in ten who were exposed to asbestos dust during the course of their work. The construction industry, including demolition, continued to receive special attention. Efforts were now being made to encourage the removal of all asbestos by specialist sub-contractors before demolition started because of the difficulty in controlling the hazard.

New cases of asbestosis and mesothelioma continue to be identified. During 1973 there were 128 deaths. The Inspectorate retained its optimism about the future prospects of controlling the disease by means of the regulations.

Since this is a long term disease, these figures reflect conditions in the past, when, in the then state of knowledge, it could not be ascertained with any certainty what levels of air contamination by asbestos dust would endanger health. There are some 1,200 factories known to be subject to the Asbestos Regulations 1969. This does not include construction sites where the application of the regulations is intermittent. Work continues on a major long-term medical and environmental survey of asbestos workers. The purpose is to learn as much about the history of exposure to different types of dust and to provide a sound basis for future assessments of preventive measures. Environmental dust measurements will be required so that any observed clinical and radiological changes, which may be discovered by medical examination can be related to the working conditions at the factories.[118]

Throughout 1974 work continued with the surveys. During the year there were 149 deaths connected with previous exposure to asbestos. Seventy-five were accepted as due to asbestosis. 139 new cases of asbestosis were recorded by the DHSS. Reports maintained that this continued to reflect conditions in past years because the latent period of this disease was such that annual figures could not yet be expected to reflect improved conditions following the new legislation in 1970. Inspectors continued to pay special attention to premises where asbestos was found on construction and demolition operations. Unsatisfactory conditions in these operations led to fourteen successful prosecutions. The high risk associated with sprayed asbestos was finally acknowledged:

> Until recently sprayed asbestos was widely used. Even minor structural alterations disturbing such coatings can give rise to concentrations liable to cause danger to health. The precautions need to be of a very high standard. Unsealed sprayed asbestos coatings have a finish that is soft, friable and subject to dusting by abrasion or attack by birds.[119]

In 1974 HM Factory Inspectorate and its Medical Branch were incorporated into the Health and Safety Executive in accordance with the Health And Safety at Work etc. Act 1974. Concern for the occupational health problems associated with asbestos dust continued at an increased pace. New legislation was set in place to exercise strict control over the use of asbestos, particularly in places where other workers or the public could be inadvertently exposed to excessive levels of the dust. Most of the concern now concentrated on the control of demolition work and the disposal of toxic waste at land-fill sites. For many this level of control had come too late to be of any benefit to them. The use of sprayed asbestos was finally prohibited by the Asbestos (Prohibitions) Regulations 1992. This was an action which on the evidence of the time could have been taken much sooner if only the risks had been properly assessed and the same control standards applied in the factories of the asbestos manufacturers had been applied when asbestos products were used in other places.

A lesson from the past

The legacy of asbestos and its inherent health risk was most clearly shown by the tragedy that took place in the small West Yorkshire town of Hebden Bridge. The Acre Mill in the Old Town was originally a woollen mill, built in 1859, ironically at a time when the Factory Inspectors were beginning to establish their place in our industrial heritage. The Dunlop Rubber Company Limited bought the mill in

1920, and in 1937 it was taken over by Cape Asbestos Limited. This company was a principal producer of crocidolite and amosite asbestos imported from South Africa. During the war years the Acre Mill was used to manufacture asbestos fibre filters for gas masks. The factory closed down in 1970 leaving behind a workforce and local population who had been exposed to the asbestos dust, either as a result of direct contact or through environmental exposure outside the premises. An increasing awareness of developing asbestos–related illnesses in this population caused the local Member of Parliament to lodge an official complaint against the Health and Safety Executive, which was now responsible for the Factory Inspectorate, for its failure to enforce the Asbestos Regulations 1969 which had superseded the original regulations first drafted in 1931. In March 1976, the Parliamentary Commissioner for Administration established that 12 per cent of the 2,200 employees had suffered an asbestos related illness.

In the essay 'Asbestos − A Failure',[19] the author provides a frank analysis on the role of the Factory Inspectorate and its responsibility for the regulation of the asbestos related industries during this period. The essay considers the critical review by the Parliamentary Commissioner for Administration who investigated the complaint and the allegation that a factor contributing to the contraction of asbestosis by a constituent was maladministration by the Factory Inspectorate. The author confirms that the Commissioner came to the conclusion that the Inspectorate did, to some degree, fall short of the standard of performance of their functions that could properly be expected of them. The essay examines some of the possible reasons for the failure of the Inspectorate, primary one of which is:

> Successful enforcement of health and safety legislation relies heavily on the inspector who visits the workplace. HM Factory Inspectorate has always operated on the assumption that inspection will be carried out by inspectors working on their own with quite limited oversight by their superiors. For this approach to be successful it is necessary for the individual inspector to be convinced that what he is doing is important and that by his work, including in appropriate cases, enforcement in the courts, he is making a significant improvement in the health and safety of people at work. He needs therefore to understand that what he is dealing with are real and significant problems and that the powers at his disposal are adequate to deal with them . . . If one assumes, and I believe it is right to do so, that almost all inspectors would have wished to reduce the amount of asbestos related disease, their apparent failure to achieve this during the period in which the Asbestos Industry Regulations were in force suggests that for some reason they had come to believe either that despite the existence of the Regulations the problem was relatively unimportant or that it was not solvable by enforcement activity.

Other reasons are advanced by the author of the essay to explain the acknowledged failure of the Inspectorate to bring this industrial disease under control in the decisive manner achieved in other cases such as lead poisoning. The complex organisation of the Inspectorate and the uncertain relationship between the District Inspectors and their specialist colleagues is given as one reason. Perhaps the main reason, which was not considered, was the inability of the Inspectors to learn from and relate to their own historic experience − an experience that was well documented in their own records.

The Health and Safety Commission reacted to this criticism by setting out to inform and alert both workers and the general public alike to the risks from exposure to asbestos. They set up an Advisory Committee on Asbestos, chaired by the Chairman of the Health and Safety Commission. A Priority of the committee was the study of the medical effects of exposure to asbestos and a review of the current hygiene standards to establish safe levels of exposure for those who might come into contact with these deadly fibres. This action was ultimately to lead to the complete prohibition of the use of this material and the strictest measures possible for its disposal. Unfortunately these measures cannot undo the legacy of the past.

CHAPTER 12

A NEW APPROACH

A time for change

The Factories Act 1937 incorporated a century of legislative experience in health and safety. It swept away the distinction between premises where mechanical power was used and those where it was not. Distinctions between textile and non-textile factories were abolished and with very few exceptions the Act was

Consequences of a workman being trapped by a conveyor belt

applied to all premises in which people were employed in manual labour or in making, repairing, altering, adapting any article for sale, whether mechanical power was used or not. It reduced the emphasis on machinery accidents by adding provisions for means of escape, maintenance of floors and stairs, and other types of non-machinery accidents. Twenty-four years later the 1937 Act was repealed and replaced by the Factories Act 1961.[58] This new Act incorporated most of the earlier provisions of the 1937 Act and remained in force throughout the remainder of the twentieth century until European Community legislation replaced many of the national requirements.

Changes in the legislation during this period were reactive in response to the many hazards, both old and new, presented by industrial development. However the annual reports continued to show an almost constant number of accidents and fatalities each year. The picture opposite provides a graphic illustration of the consequences of one of these accidents. It first appeared in an edition of accidents in 1958. Table 12.1 is a summary of all industry related accidents over three-year periods, recorded between 1920 and 1956. The statisticians of that period were able to qualify these figures by showing they represented a reduction in the incidence rate and hence the risk — but the hazards remained and our industrial society continued to make an unacceptably high demand upon those exposed to its many dangers.

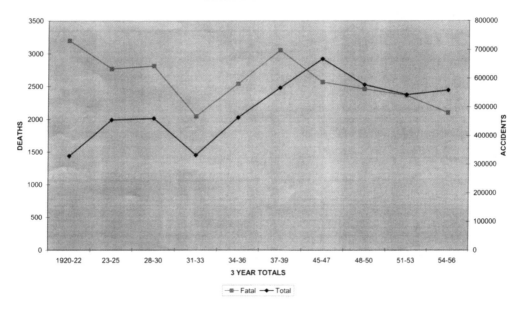

Accident statistics 1920 to 1956

Table 12.1 Summary of all accidents from 1920–1956

Safety of plant and machinery was well covered in the 1961 Act but for many reasons the legislation was found to be outdated and still in need of change. Fatal and non-fatal accidents such as those illustrated in Table 12.1 were the result of both traditional and new industries creating risks to those who came into contact with dangerous parts. These accident statistics were accumulated by the Factory Inspectorate and recorded in the annual reports of successive Chief Inspectors of Factories. The returns included accidents caused by specific types of machinery such as prime movers, transmission machinery, power presses, lifting equipment and woodworking machinery. Tables 12.2 and 12.3, which are derived from the data in Table 7.5, illustrate the total number of accidents and deaths, over 2 year periods, associated with these five categories of machinery and equipment. The most significant point to note is the high and relatively constant accident rate for lifting equipment and the steady decline in accidents caused by transmission machinery.

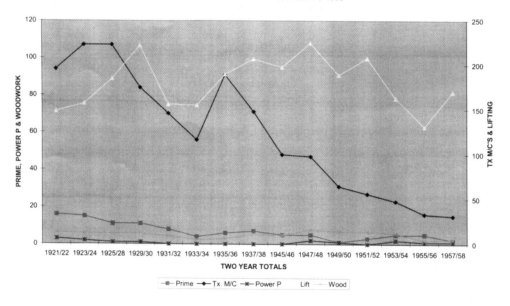

Table 12.2 Summary of fatal machinery accidents from 1921–1958

Opposite: Table 12.3 Summary of machinery accidents from 1921–1958

Accidents - Machines 1921 to 1958

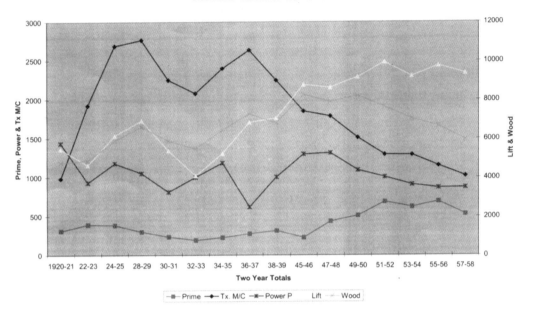

The enactment of the 1961 Act brought about the introduction of many new codes of regulations to cover a wide range of industrial processes and machinery, which required a higher level of technical requirements. Many of these regulations were derived from the accumulated experience of the Inspectorate over the previous forty years, much of which was due to the specialist input from the Technical Branches of the Department. The following regulations, which were made under this legislation were particularly relevant to this expertise:

Construction (Lifting Operations) Regulations 1961
Examination of Steam Boilers Regulations 1964
Power Presses Regulations 1965
Abrasive Wheels Regulations 1970
Highly Flammable Liquids and LPG Regulations 1972
Woodworking Regulations 1974

Much of this machinery and equipment was subject to prescribed statutory examination and reporting carried out by competent persons such as those employed by the engineering insurance companies. With regard to the examination of plant, one of the more important pieces of legislation to be written was the Examination of Steam Boiler Regulations of 1964.[59] These regulations were derived from the recommendations of the Honeyman Committee[60] appointed to look at the statutory examination requirements for steam boilers in the light of modern technical developments. These developments brought into question, the validity of the statutory fourteen-month examination. This examination period was derived from a time when factories and mills shut down for annual holidays, thus making their steam plant available for detailed internal examination. Very large steam generation units, installed in power stations and chemical installations operating continuously, made unworkable the legal requirements that were formulated in a different industrial era. The technical committee took evidence from a wide body of opinion and was able to make recommendations which relaxed the periodicity of examination for some types of steam plant without compromising the overall safety objectives achieved by fixed statutory examination periods.

The new regulations were another step in addressing the needs of the changing technology of the time. They increased the periods between examinations for certain types of boilers but specified the need for a more thorough and comprehensive inspection. Additionally, the Chief Inspector of Factories had the power to grant Certificates of Exemptions for certain types of plant where it was impracticable or unnecessary to carry out examination, provided safety was not compromised by such exemptions. Examination of boilers by competent persons was recognised as an important part of the new regulations. The meaning of competent person was established in the decided legal case (Brazier v. Skipton Rock Co Ltd)[61] and formed an important part of future legislation:

> A competent person should have such practical and theoretical knowledge and actual experience of the type of machinery or plant which he has to examine as will enable him to detect defects or weakness which it is the purpose of the examination to discover and to assess their importance in relation to the strength and functions of the particular machinery and plant.

The Examination of Steam Boiler Regulations were finally revoked, thirty years later, in 1994 by the Pressure Systems and Transportable Gas Containers Regulations[62] made under the Health and Safety at Work etc. Act 1974. Whereas legislation for pressure equipment changed and adapted to the new approach under the Health and Safety at Work Act, similar progress did not take place with the other equipment covered by the 1961 Act. There was an attempt to introduce new lifting equipment legislation with a proposal for Lifting Gear (Testing & Using) Regulations but the consultation with industry was not successful and the proposal was dropped and is now covered by European Community legislation.

The Health and Safety at Work etc. Act 1974

Although the various Factories Acts up to 1961 were successful in reducing, or at least containing industrial accidents and improving the level of occupational health and welfare, the underlying statistics were a cause for concern. Various committees of inquiry were appointed during the post-war years to look at the legislation and make recommendations to the government. This culminated in the appointment of the Robens Committee in 1970. This appointment arose from a sense that an acceptable level of health and safety at work had not been reached, and if standards were to be raised new methods and a new approach were needed. The Robens Committee[63] completed its work two years later. Recommendations made by the committee were based on the philosophy that those who created the risks should be primarily responsible for their identification and removal. The Committee proposed:

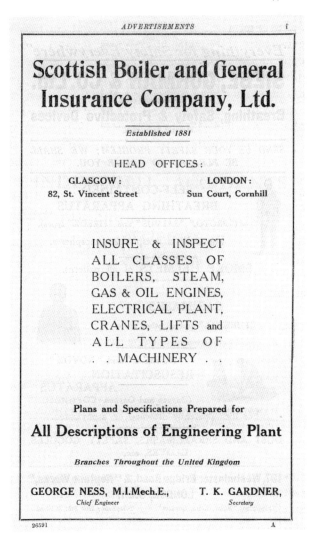

Typical advertisement by an
Engineering Insurance Company

A comprehensive and orderly set of revised provisions under a new enabling Act. The new Act should contain a clear statement of the basic principles of safety responsibility. It should be supported by regulations and by non-statutory codes of practice, with emphasis on the latter. A determined effort should be made to revise, harmonise and update the existing large body of detailed statutory regulations, to simplify their style and to reduce their number. The scope of the new legislation should extend to employers, employees and self-employed.

Parliament accepted the recommendations and implemented them under the Health and Safety at Work etc. Act 1974.[64] The Health and Safety Commission was set up to oversee the new legislation and appointed representatives from industry, the trades unions and other relevant bodies to ensure a comprehensive oversight of the new Act. Executive functions were delegated to a new organisation, the Health and Safety Executive, which included the Factory Inspectorate as an important part of its operational and policy functions. The new Act was an enabling Act, which gave powers to the Secretary of State to make legislation in the form of regulations that had to be laid before Parliament before becoming law. Power was given also to issue Approved Codes of Practice on relevant matters and to be able to approve national standards if they were relevant to health and safety. The intention was to replace the existing legislation such as the Factories Act with regulations supported by codes of practice and other guidance.

The new Act was drafted so that the rigid statutory demands of earlier legislation would be replaced by more general duties placed upon those carrying out activities where work was involved. These general duties were prescribed in Sections 1 to 9 of the Act and were expressed in terms of compliance 'so far as is reasonably practicable'. Gradually the legislation was moved from a policy of procedural certainty to one where compliance with the law required a degree of judgement on the part of those covered by the Act. The requirement to carry out one's duties so far as is reasonably practicable caused some uncertainty. Reasonably practicable had a legal meaning derived from the legal case involving (Edwards *v.* National Coal Board) where the Court of Appeal decided:

> Reasonably practicable is a narrower term than physically possible and seems to imply that a computation must be made in which the quantum of risk is placed on one scale and the sacrifice involved in the measures necessary to avert the risk (whether in money, time or trouble) is placed on the other, and that, if it be shown that there is a gross disproportion between them – the risk being insignificant in relation to the sacrifice – the defendants discharge the onus on them. Moreover this computation falls to be made by the owner at a point of time anterior to the accident. The questions he has to answer are; what measures are necessary and sufficient to prevent any breach and are these measures reasonably practicable.[65]

Section 6 of the Act placed duties on persons who design, manufacture or supply articles for use at work to ensure so far as is reasonably practicable that the articles will be safe and without risk to health at all times. There was an additional duty to carry out testing and research to eliminate or minimise any risk to which the article may give rise. Similarly, Section 2 of the Act placed duties on employers to ensure so far as is reasonably practicable the health, welfare and safety of their employees at work by the provision and maintenance of plant and systems of work that are safe and without risk to health.

Health and safety legislation in the form of the 1974 Act moved away from a reactive solution to a particular problem to a more general duty of care derived from a careful consideration of the likely risks to health and safety if reasonable actions were not taken by those who in the process of their work activity created that risk. What is more, that duty of care extended to members of the public and others who might be inadvertently exposed to the risk so generated. Much of the legislation, which was enacted under the Act was based upon an assessment of hazards and a requirement to minimise the risk of that hazard being realised. Risk assessment was the new approach to an acceptable policy of informed judgement, and the political tool in obtaining general and public acceptance of potentially hazardous activities. In time it may be possible to judge whether the philosophy of risk assessment had more success and permanence than the traditional legislative framework of reacting to specific problems of health and safety. There is a delicate balance between a decision for positive action and no action required on grounds of a perceived risk being tolerable to the public. The general acceptance of the high accident rates such as that on the roads compared with rail or aircraft accidents provide good examples of different levels of tolerable risks.

Flixborough and its influence on legislation

On 1st June 1974 the Flixborough Works of Nypro (UK) Limited was destroyed by an explosion of warlike dimensions. Twenty-eight employees were killed and thirty-six workers suffered injury in the incident. Outside the works fifty-three members of the public were injured and there was widespread property damage. The Health and Safety at Work etc. Act was enacted by Parliament on 31 July, exactly two months after the accident. So it came about that two significant events came together with uncanny timing. These events had a major effect on the course of health and safety legislation. On 27 June the Secretary of State for Employment, Mr Michael Foot, directed that a formal investigation be held into the causes of the Flixborough accident. After hearing evidence from many witnesses and experts, the Court of Inquiry produced a report, [66] which detailed the lessons to be learned from the disaster. In addition it recommended that a special committee should be set up to consider specific matters that were beyond the scope of the Inquiry. This new approach to health and safety based upon hazard analysis and risk assessment, which found its origins in the nuclear industry, was subject to public scrutiny at this Inquiry.

The Court of Inquiry accepted the view presented by one witness that:

Hazard analysis recognises that hazard cannot be entirely eliminated and that it is necessary to concentrate resources on those risks which exceed a specific value.

In the final report the Court of Inquiry concluded:

We agree with this statement which accords with reality. No plant can be made absolutely safe anymore than a car, aeroplane, or home can be made absolutely safe. It is important that this is recognised for if it is not, plant, which complies with whatever may be the requirements of the day tends to be regarded as absolutely safe and the measures of alertness to risk is thereby reduced.

When Mr Marshall refers to risks exceeding a specific value we understand him to refer to risks which exceed what at a given time is regarded as socially tolerable, for what is or is not acceptable depends in the end upon current social tolerance and what is regarded as tolerable at one time may be regarded as intolerable at another.

It would be both wrong and impossible for us to attempt to assess the current level of social tolerance in relation to risk within the chemical industry but the evidence before us covered a very wide field and has, we believe, enabled us to draw attention to certain objectives which could be usefully considered by the industry itself and by other bodies concerned with safety.

These opinions coupled with the new approach in the Health and Safety at Work etc. Act with its legal provisions derived from the interpretation of what is reasonably practicable combined to set in motion an entirely new thinking on the control of industrial hazards. Further to this, by defining tolerability in terms of social values it introduced the possibility of these values changing with time either to be more tolerant or less tolerant of a particular risk. The more orthodox approach to health and safety was not completely ignored because the Court of Inquiry found in the evidence presented that the chemical industry was not subject to the same type of statutory inspection as applied in factories and made the following recommendations:

> Pressure systems containing hazardous materials should be subject to inspection and test by a person recognised by the appropriate authority as competent.

> Existing regulations relating to steam boilers which do not apply to pressure systems containing hazardous materials should be extended so as to apply to such systems . . . consideration should be given to whether the inherent nature of the material requires greater control than applies in the case of steam boilers.

In effect these recommendations were endorsing the application of the statutory requirements of the old Factories Acts upon an industry that had avoided this type of prescriptive legislation. The Court of Inquiry identified other matters, which were to be referred to a special committee. This was accepted and one of the first duties placed upon the Health and Safety Commission was to appoint the Advisory Committee on Major Hazards to examine the following topics:

> Plant Layout and Construction — to investigate conditions prevailing at other plants, not only in the United Kingdom but elsewhere.
> Siting of Plant — it must depend on both the risk of a disaster and the size and nature of the disaster envisaged should the risk materialise.
> Planning Procedures — to consider co-ordination between the planning authorities and HSE.
> Emergency Arrangements — to consider a disaster plan for the co-ordination of rescue, fire fighting, police and medical services.
> Licensing Storage — to review the existing regulations.

The Advisory Committee comprised representatives from a wide range of interests. It produced three major reports between 1976 and 1984,[67, 68, 69] with recommendations that changed the United Kingdom approach to safety, particularly in the area of major hazards in the chemical industry.

In the First Report[67] it was noted that the pressures for greater efficiency led to the growth of capital intensive new plants such that there was a significant increase in the number of people who could be endangered at any one time should anything go wrong. Moreover it stated that the pace of change associated with modern technology allowed less opportunity for learning by trial and error. The 'Unrelenting Machine' was now recognized as an installation comprising a complex of systems in which those at risk were now an integral part of that installation, where the traditional protection of safeguarding by position no longer applied. A number of recommendations arose from this:

a) A Major Hazards Assessment Unit should be set up in HSE.

b) A scheme for the notification of certain companies should be set up whereby these 'Notifiable Installations' should be required to submit to HSE a survey of the plant and the procedures and methods to deal with the hazards.

c) A legal requirement should be made for consultation between HSE and Local Authorities prior to granting planning permission for certain developments.

By the time the committee issued its Third Report[69] in 1984 the HSE Assessment Unit had been formed and the following safety legislation was in place:

Notification of Installations Handling Hazardous Substances Regulations 1982 (SI 1982 No.1357).

Control of Industrial Major Accident Hazards Regulations 1984 (SI 1984 No.1902) (implementing EC Directive on Major Accident Hazards 82/501/EEC)

Both the Flixborough report and the reports of the Advisory Committee recognised the importance of the traditional statutory requirements that applied to steam plant and the need for engineering judgment in dealing with these matters. But apart from the Pressure Systems and Transportable Gas Containers Regulations 1989 ,which only covered the hazard of stored energy, the general recommendations in this field were largely ignored in preference to the risk based philosophy with emphasis on the user deciding what was reasonably practicable. The recommendations of the Advisory Committee in respect of a possible licensing scheme for hazardous installations included the following recommendations for acceptance of the design and operation of pressure systems:

Management should ensure that an appropriate inspection system is operated during fabrication and construction to check that the standards set by the management design authority are being met. This inspection system may be in the same management organisation as the design authority, or it may be appropriate to use one of the engineering insurance companies or other approved inspection agency. The inspection system should not be part of the operating authority . . .

The operating authority should also provide and enforce the use of a code which, ensures the continuing safety of the pressure system by a regular inspection of the equipment and the safety devices which are provided to protect that equipment . . . The code should specify rules, which guarantee the independence of the inspection system from the operating authority, either by appointing external inspection authorities, or by making satisfactory arrangements for the in-house inspection authority to be responsible to a senior member of the organisation who has the design authority within his charge.

Engineering control of the operations, as part of the licensing conditions, would have applied to all pressure systems containing hazardous substances if the Committee recommendations had been carried out. But licensing of notifiable installations did not happen and the legislation that was eventually introduced was less demanding, being limited to such requirements as notification of hazardous installations, provision of information and the submission of safety reports to the Executive by the operator.

The first draft regulations for pressure systems were initially very comprehensive but industry objected when the proposals for the new regulations were sent out for public comment. In the end the regulations were limited to the hazard of stored energy from certain relevant fluids and certainly not as intended by the committees set up to consider these important safety matters. Industries acceptance of new regulations covering pressure systems was only possible because of the Flixborough explosion in 1974. This incident alerted the public to the danger from industrial plant containing hazardous materials. It took two public consultation documents in 1978 and 1984 and approximately eighteen years for the recommendations to be acceptable to industry and to appear as the Pressure Systems and Transportable Gas Containers Regulations 1989. It is unlikely that the United Kingdom will, at a national level, generate such a radical new approach to legislation on such important matters. All of our future health and safety legislation will be in accord with the consensus of our European partners. Although independent regulatory inspection and control of plant and equipment was a well-established tradition in other Member States, this approach was not incorporated into European Directives. The European-led legislation was influenced by considerations of hazard and risk and the setting of different levels of conformity to the law in accordance with that risk.

Application of risk assessment principles

Shortly after the Flixborough report was published in 1975, an opportunity arose for the Health and Safety Commission to put the recommendations on risk assessment to the test as part of a public inquiry into a planning application to extend the petrochemical installation at Canvey Island on the north side of the River Thames. This was the first time that a non-nuclear regulatory authority was prepared to apply these procedures as part of its decision making process. In March 1976 the Commission agreed to a request from the Secretary of State for Employment, the Rt Hon. Albert Booth MP, to carry out an investigation of the risks to health and safety associated with various installations, both existing and proposed, on Canvey Island and Thurrock. The HSE carried out the investigation and were assisted by the United Kingdom Atomic Energy Authority's Safety and Reliability Directorate. Two years of investigation and theoretical analysis followed before the final report in two parts was completed at a cost of £400,000. Part 1[70] summarised the investigation and the recommendations to ensure improvement of the installations. Part 2[71] contained the detailed technical report of the investigation teams based upon the information obtained about the existing installations and the proposed developments that were subject to the planning application.

The report made a number of recommendations for improvements to the existing and proposed new developments that would enable the application to proceed without objection from the Health and Safety Executive. This decision was supported by a detailed numerical assessment of risk. This was certainly a new approach and one that was acknowledged in the report as likely to be controversial. The assessment of risk was expressed firstly as the annual chance of an individual being killed (individual risk), and secondly as the chance of killing more than 10 people at a time (societal risk). The report concluded that the chances of a person in Canvey Island or Stanford-le-Hope being killed as a consequence of a major accident at an existing installation was about the same as the average risk of a person in the twenty-five to thirty-four age group in this country dying from natural causes. It was concluded that this level of risk, five chances in 10,000 per year, would be reduced by about two thirds if all the improvements asked for in the report were adopted. In 1981 a reassessment of the 1978 report was carried out by the Safety and Reliability Directorate on behalf of the HSE in which it was concluded that the estimated individual risk could now be assessed as 0.35 chances in 10,000 a year. The reports, which were first of their kind in the United Kingdom, greatly influenced the HSE policy on health and safety issues. The quantitative analysis of risk based upon theory and technical judgement provided a powerful political tool to counter the more emotional response likely to come from public objection to such developments. Risk assessment became an integral part of subsequent United Kingdom and European legislation and is now a necessary prerequisite to the simplest of activities where there might be some perception of a risk of injury or ill health.

The European Economic Community

The movement towards European integration started with a Treaty in 1951,[72] which established the European Coal and Steel Community. This was followed by two Treaties, which set up the European Atomic Energy Community and the European Economic Community, both signed in Rome in 1957. The United Kingdom joined on 1 January 1973 and implemented the Treaties by enacting the European Communities Act 1972. [73] This established the deference of United Kingdom law to Community law and its law-making and judicial institutions. In 1986 all the member states signed the Single European Act, which amended the Treaty of Rome by incorporating a new Article 118A to the Treaty, which permitted the Community to introduce minimum standards for the health and safety of workers. This meant that the standard setting was by Directives decided on the qualified majority voting of the member states. Although the Directives were binding, the means of implementing each Directive into law was left to the Member State.

Prior to 1987 only six Directives on health and safety at work had been made by the European Commission and incorporated into United Kingdom law. These Directives were issued under Article 100 of the Treaty of Rome, which required unanimity in the voting. Perhaps the most significant of these was the Biological Agents Directive 80/1107/EEC, which was implemented into UK legislation as the Control of Substances Hazardous to Health Regulations 1988 (COSHH). Other Directives concerned with the control of noise and the control of asbestos originated from this period. Article 100 was also the basis for Directives where the protection of workers was incidental to the main purpose. Such Directives included the Classification, Labelling and Notification of Dangerous Substances

Directive 67/548/EEC, implemented in the United Kingdom by the Classification, Packaging and Labelling of Dangerous Substances Regulations 1984 and the Major Accident Hazards Directive 82/501/EEC, implemented in the UK by the Control of Industrial Major Accident Hazards Regulations 1984 (CIMAH).

There were a number of other important Regulations in place within UK law, which originated from our implementation of the European Communities Act 1972.

The Single European Act and Article 118A Directives

The Single European Act[74] in 1986 and the new Social Dimension gave effect to a European Commission initiative in the health and safety at work sector. A Third Programme of legislation was proposed by the Commission, which included fifteen new Directives. From this initiative came the principal Directives which culminated in a package of new UK Regulations in 1993. The most important of these Directives was the Framework Directive 89/391/EEC, which required that employers take preventative and protective measures to avoid risks; to evaluate risks which cannot be avoided; to combat risks at source; to adapt the work to the individual; to adapt to technical progress; to replace the dangerous by the less dangerous and develop a coherent overall policy towards the health and safety of the workforce. The Framework Directive was followed by six 'Daughter Directives', which had to be implemented by 1 January 1993:

> The Workplace Directive 89/654/EEC
> The Work Equipment Directive 89/655/EEC
> The Personal Protective Equipment Directive 89/656/EEC
> The Manual Handling of Heavy Loads Directive 90/269/EEC
> The Display Screen Equipment Directive 90/270/EEC
> The Carcinogens at Work Directive 90/394/EEC.

The Framework Directive was incorporated into UK law by the Management of Health and Safety at Work Regulations (S.I. 1992 No.2051) and the six daughter Directives by the following Regulations, each supported by Codes of Practice:

> Workplace (Health, Safety and Welfare) Regulations (S.I. 1992 No.3004)
> Provision and Use of Work Equipment Regulations (S.I. 1992 No.2932)
> Personal Protective Equipment at Work Regulations (S.I. 1992 No.2966)
> Manual Handling Operations Regulations (S.I. 1992 No.2793)
> Health and Safety (Display Screen Equipment) Regulations (S.I. 1992 No.2792)
> Control of Substances Hazardous to Health (Amendment) Regulations (S.I. 1992 No.2382 and S.I. 1992 No.2966)
> Control of Asbestos at Work Regulations (S.I. 1992 No.3068)

The Framework Directive, the Workplace Directive and the Work Equipment Directive were relevant to the safety of plant and machinery. Risk assessment was a requirement of the Framework Directive and the consequent UK Management of

Health and Safety at Work Regulations 1992. These Regulations placed a number of duties on employers and self-employed persons:

a) To make and review a suitable and sufficient assessment of risks to health and safety of employees and non-employees affected by their work
b) To carry out effective planning, organisation and review of preventive and protective measures
c) To provide health surveillance for employees, appropriate to risk
d) To provide adequate health and safety training
e) To appoint one or more competent persons.

The Workplace Directive and the implementing UK Regulations replaced extensive parts of the Factories Act and the Offices Shops and Railways Premises Act. They dealt with the structure and layout of workplaces as they affected workers and the facilities provided for workers. A general requirement was that the workplace and the equipment, devices and systems were to be maintained in an efficient state, in efficient working order and in good repair, with respect to health, safety and welfare. The Provision and Use of Work Equipment Regulations 1992 was equally important in setting new requirements and standards for health and safety. Duties were imposed upon employers to provide and maintain suitable work equipment that conformed to standards specified in other legislation. These duties included:

a) To provide adequate training, information and consultation
b) To restrict equipment likely to carry risk to trained employees
c) To protect against dangerous parts and other hazards
d) To provide visible controls, safety markings and warning devices for equipment
e) To provide isolation from energy sources
f) To protect against risks while maintaining equipment.

Reference to standards in other legislation referred not only to safe systems of work but to the safety and initial integrity of plant and machinery and provided an important link with the Directives made under Article 100A of the Single European Act.

The Single European Act & Article 100A Directives

In addition to Directives made under Article 118A, other Directives emanated from Article 100A of the Treaty of Rome which were directed to the harmonisation of laws to facilitate the establishment of the internal European Market. The Machinery Directive 89/392/EEC as amended by 91/369 covered machinery, and was implemented into UK law by the Supply of Machinery (Safety) Regulations 1992 (S.I. 1992 No.3073).

This process of drafting Directives which covered the supply of equipment under Article 100A and the safe use of that equipment under Article 118A of the Treaty of Rome was implemented over a long period of time to allow the member states to adapt to the new unified requirements, which had to be implemented through their own national legislation. It remains to be seen whether the political intent of Article

100A will survive the other challenges that face the European Community. These Directives broke new ground so far as setting standards for plant and machinery were concerned. The Machinery Directive defined only essential health and safety requirements of a general nature which were supplemented by a number of more specific requirements for certain categories of machinery.

In order to help manufacturers prove conformity to the essential requirements it was policy to develop technical standards harmonised at European level for the prevention of risk arising out of the design and construction of machinery. These standards were drawn up by private non-government bodies and had no binding legal status. For this purpose the European Committee for Standardisation (CEN) and the European Committee for Electromechanical Standardisation (CENELEC) were the bodies recognised as competent to adopt harmonised standards. A harmonised standard was defined as a technical specification adopted by either or both of these bodies on the basis of a remit from the European Commission for the provision of information in the field of technical standards. Conformity to a harmonised standard created a presumption of conformity to the relevant essential requirements of the Directive.

Manufacturers retained responsibility for certifying the conformity of their machinery to the relevant essential requirements, with or without reference to a harmonised standard. It was left to the discretion of the manufacturer whether he felt the need to have his products examined or certified by a third party. However, for certain types of machinery with a higher risk factor, a stricter type of certification was required. In such cases the manufacturer was required to complete a technical file and send it to an Approved Body for verification purposes, alternatively the manufacturer could opt to send a sample of the machine for type examination by the Approved Body.

Approved bodies and competent persons

Directives made under Article 100A were intended to remove technical barriers to trade. Primary responsibility was placed upon the manufacturer to show that his equipment conformed to the appropriate Directive. In those cases where an external organisation was used in the conformity assessment procedures that organisation was appointed by a National Authority and had to satisfy minimum criteria of competence and integrity. Such organisations were known as Approved Bodies in United Kingdom legislation and Notified Bodies in the 100A Directives. The Supply of Machinery (Safety) Regulations 1992, which implemented the Machinery Directive is the best example of how the conformity procedures would work in practice.

Responsibility for conformity assessment, for most types of machinery, was established by the appointment of a Responsible Person who was the manufacturer of the machine or his authorised representative in the Community. The Responsible Person was required to draw up a technical file and carry out the appropriate conformity assessment procedures required by the legislation so that an EC Declaration of Conformity and other necessary certification could be provided to show that the machine satisfied all the relevant essential requirements and, where appropriate was in accordance with the harmonised standard.

In the case of machinery that posed special hazards, Annex IV of the Directive and the Regulations required the additional services of an Approved Body. In this case the Responsible Person was required to draw up the technical file and send it or a sample of the machine to the Approved Body to obtain verification that the machine in question conformed to the relevant standards and essential requirements. A Certificate of Adequacy or an EC Type Examination, as appropriate, was then obtained from the Approved Body so that the manufacturer could issue an EC Declaration of Conformity. The type of machinery listed in Annex IV was limited and certainly did not cover all hazards likely to cause accidents when the machines were being used. Later draft Directives such as the Pressure Equipment Directive 97/23/EC had more complex requirements ranging from the straight forward manufacturer declaration of conformity to assessment procedures based upon Quality Management principles and standards. Notified Bodies could include both third party and second party interests.

Article 118A Directives and Regulations were less specific about this matter. For example, the Management of Health and Safety at Work Regulations, required every employer to appoint one or more competent persons to assist in undertaking the measures he needed to take to comply with the duties imposed upon him by the statutory provisions. In such cases the person could be regarded as competent where he or she had sufficient training and experience or knowledge and other qualities to enable him or her to properly assist in undertaking the measures referred to. In most cases such a person would be an employee of the company.

The Provision and Use of Work Equipment Regulations covered all types of risk that workers were likely to be exposed to during the course of their employment. This covered the field of the Factories Act 1961 and the many sets of safety regulations implemented in the UK over the years. The new regulations repealed parts of the Factory Act concerning the secure fencing of machinery. Despite the relevance of the new Regulations to safety there was no reference to periodic inspection or examination by competent persons

Significantly, the Directive did have provisions for laying down additional minimum requirements to provide protection against specified hazards. The European Commission made a proposal to amend the Directive in respect of certain types of equipment where deterioration in service could pose special risks. The proposal identified the need for initial inspection and in-service inspection for equipment such as pressure vessels and mobile lifting equipment:

> The employer shall ensure that where the safety of work equipment depends on the installation conditions, it shall be subject to initial inspection after installation and before being taken into service for the first time.
>
> The employer shall ensure that work equipment exposed to conditions causing deterioration, which are liable to result in dangerous situations are subject to periodic inspections.
>
> The employer shall draw up or have drawn up a plan of inspection of work equipment based upon the intended use and the plan shall determine the type and frequency of the inspection.
>
> The Member States shall specify the criteria governing the competency of competent persons for drawing up the inspection plan and carrying out the inspections.

In the light of experience and the lessons of history, it is evident that these requirements would provide an important and essential contribution to health and safety. Unfortunately the proposal was dropped and one must question on what basis such an important decision was reached when so many other untried changes have been introduced in recent times. Ironically, the principles of inspection by competent persons, which were rejected in areas where they have proved their value, were adopted in other legislation. The Carcinogens at Work Directive implemented in the UK by the Control of Substances Hazardous to Health Regulations 1988 was a case in point. Under these regulations an employer was not permitted to carry out any work that was liable to expose any employee to any substance hazardous to health unless he made a suitable and sufficient assessment of the risks and of the steps that needed to be taken to meet the requirements of the Regulations. Duties were placed upon the employer for the maintenance, examination and test of control equipment to ensure it was in efficient working order and in good repair. Where engineering controls were provided the employer had to ensure that thorough examinations and tests were carried out by competent persons, and in the case of local exhaust ventilation plant this was to be done at least once every fourteen months.

Assessment of risk on the part of those who created the risk had become the accepted philosophy and those with an interest in that approach were able to influence the thinking of those in authority. This new approach formed the basis of many pieces of new legislation coming from the European Community. The future developments in this field were well summarised as follows:

> The history of health and safety legislation has by no means reached an end. It has however entered a new and different stage. The European revolution is characterised by universality of application and a focus on the nature of the risk (rather than on the nature of the workplace or the work process). Millions of workers previously outside the scope of the legislation will receive statutory health and safety protection and employers will be stimulated to assess and act upon the risks of all their operations. The duty to carry out risk assessment will become a prominent feature of forseeability in negligence actions. Diminution in the scale of the horror of injury, ill health or death at work is to be anticipated. [75]

To some extent this is a great social experiment that has yet to run its course. Since Flixborough we have seen an escalation of similar and much worse disasters at home and abroad, which in earlier times might have resulted in the imposition of stringent controls over the industries involved. In the occupational health field we see a much greater degree of control being imposed now because of a heightened awareness of the dangers of hazardous substances. The control over the use of asbestos in latter part of the twentieth century contrasts starkly with the attitude that existed towards the known hazard of asbestos at a time when the machinery hazards were being so tightly regulated. If regulatory authorities had adopted the same measures to control the use of asbestos in the 1950s, the burden of ill health and financial loss by the insurance industry in the 1990s would have been significantly less. This confirms that the tolerability of risks can be changed and in consequence the extent to which it is considered necessary to control the risks by legislation can also be changed.

Since risk is seen as being unavoidable in an industrial society and since there is no absolute duty to completely eliminate the risk, then when that risk materialises as an accident or dangerous occurrence, a price has to be paid. Society at large should be best placed to judge whether such risks are truly tolerable or whether measures are necessary to effect change.

NB

This chapter provides a commentary on the new approach to national and European legislation as it was towards the end of the twentieth century. Since then, significant changes have been made to European Directives with corresponding amendments to our national regulations. Current legislation needs to be understood by those who have a legal and moral responsibility for the protection of persons at work and members of the public who may be affected by that work.

List of references

1. A. URE. *The Philosophy of Manufactures* (1835).
2. M.T. SADLER. Factory Statistics (1836).
3. F. ENGELS. *The Conditions of the Working Class in England in 1844.*
4. E. BAINES. *History of the Cotton manufacture in Great Britain.* 1835.
5. GULIELMI IV. REGIS. CAP CIII. 29th August 1833. An Act to regulate the labour of Children and Young Persons in the Mills and Factories of the United Kingdom.
6. TREVOR MAY. *An Economic and Social History of Britain 1760−1970.* ISBN 0 582 35280 0
7. L. HORNER. Report of 4 January 1841.
8. L. HORNER. Report of 1 May 1844.
9. VICTORIAE REGINAE. CAP XV. 6th June 1844. An Act to amend the Laws relating to Labour in Factories.
10. L. HORNER. Report of 30 April 1845 Part Papers (1845) XXV, p246.
11. VICTORIAE REGINAE. CAP XXXIII. 30th June 1856.
An Act for the further Amendment of theLaws relating to Labour in Factories.
12. Factory Inspectors celebrate 150 years. Health & Safety at Work. Aug 1983.
13. VICTORIAE REGINAE. CAP III. 15 August 1867. An Act for the Extension of the Factory Acts.
14. Factory and Workshop Act 1878. (41 VICT. CH 16).
15. Factory and Workshop Act 1891.
16. The Employers Liability Act 1880.
17. Factory and Workshop Act 1901. (1 EDW. 7 CH22).
18. The Health and Morals of Apprentices Act, 1802.
19. *Her Majesty's Inspectors of Factories 1833-1983.* HMSO 1983. ISBN 0118837117.
20. THOMAS, Maurice Wallace. *The Early Factory Legislation: A Study in Legislative and Administrative Evaluation.* 1948.
21. Joint Report 31 Dec 1846. Part papers (1847) XV p486.
22. Joint Report 30 April 1846. Part papers (1846) XX p625.
23. MARTINEAU, H. *The Factory Controversy 1855.*
24. Report of the Departmental Committee on Accidents in Places under the Factories and Workshop Acts. (Cd 5535).
25. *A Brief History of HM Factory Inspectorate.* HMSO 1980. ISBN 0 7176 0046 7.
26. An Act for the further Amendment of the Laws relating to Labour in Factories. 30 June 1856.
27. DICKENSON H.W. & JENKINS R. *James Watt and the Steam Engine.* ISBN 0 903485 92 3.
28. MARX K. *Das Fapital* (A Critique of Political Economy) Vol.1. p223.
29. Report of the Departmental Committee on Accidents in Places under the Factories and Workshop Acts. 1911 (Cd 5535).
30. Effect of the Second Year of the War on Industrial Employment of Women and Girls.
31. Effect of the Third Year of the War on Industrial Employment of Women and Girls.

32. The Factories and Workshop Act 1879.
33. *The Engineer.* 20 June 1856.
34. *The Engineer.* 18 April 1856.
35. Explosion of Petrol in a Road Tank Car at Works of Shell Mex Ltd. on 28 August 1924.
36. Explosion of New Oil Tank at Works of the Medway Oil and Storage Co. Ltd. on 14 January 1925.
37. Report of the Home Office Departmental Committee on the Factory Inspectorate
  − 18 August 1928.
38. Building Regulations 1926 (SR & O 1926 No.738) & Docks Regulations 1925 (SR & O No.231).
39. Report on Fencing and Safety Precautions for Transmission Machinery by
    W S Smith HM Inspector of Dangerous Trades − 1913.
40. Form 1811. Report of an Inquiry into the methods of Fastening Belts Used for Power
    Transmission in Factories by, E L Macklin OBE. October 1925.
41. Factory Form 418. Memorandum on lifting by G S Taylor OBE. Home Office 1929.
42. Engineering Research Special Report No.3. The causes of failure of wrought
  iron chain and cable by H J Gough & A J Murray. NPL 1928.
43. Power Press Safety Code. Fifth Report of Joint Standing Committee on Safety
  in the Use of Power Presses,1965. HMSO SBN 11 360370 3.
44. Staffing and Organisation of the Factory Inspectorate, October 1956. HMSO Cmd.9879.
45. T C SMOUT. *History of the Scottish People. 1580−1830.*
46. A MASSIE. *Glasgow: Portraits of a City.* ISBN 0 7126 2054 0.
47. S. CHECKLAND. *Dictionary of Scottish Business Biography.*
48. *The Engineer* − Highlights of 120 Years. November 1976
49. An Act for the Extension of the Factory Acts, 1867. 30 & 31 Vict, Ch. 103.
50. Factory and Workshop Act, 1878. 41 Vict, Ch. 16.
51. Factory and Workshop Act, 1891. 54 & 55 Vict, Chapter 75.
52. A Report on an Explosion at the Synthetic Chemical Works of Messrs W J Bush & Co. Mitcham.
53. Guglielmo Marconi. The Marconi Company Limited 1974.
54. Report of the Committee Appointed by Surgeon General H S Cumming. United Public Health
    Service 17/01/26.
55. *The Guardian* 13/12/94.
56. *The Guardian* 25/11/94.
57. Heat Insulation. William Kenyon & Sons Limited.
58. The Factories Act 1961 (9 & 10 Eliz. 2, c. 34).
59. Examination of Steam Boiler Regulations 1964 (SI 1964 No.781).
60. Report of the Advisory Committee on the Examination of Steam Boilers in Industry.
  Cmnd. 1173. HMSO Reprinted 1962.
61. Brazier v Skipton Rock Co Ltd.
62. Pressure Systems and Transportable Gas Containers Regulations 1989 (SI 1989/269).
63. Report of the Robens Committee on Safety and health at Work (1972 Cmnd. 5034).
64. Health and Safety at Work Etc. Act 1974 (1974 c. 37).
65. Edwards v National Coal Board. C.A. 1949, 1 AER 743.
66. The Flixborough Disaster. Report of the Court of Inquiry. HMSO 1975. ISBN 0 11 361075 0.
67. Advisory Committee on Major Hazards. First Report. HMSO 1976. ISBN 0 11 80884 2.
68. Advisory Committee on Major Hazards. Second Report. HMSO 1979. ISBN 0 11 83299 9.
69. The Control of Major Hazards. Advisory Committee on Major Hazards.
  Third Report. HMSO 1984. ISBN 0 11 3753 2.
70. Canvey. Summary of an investigation of potential hazards from operations in the
  Canvey Island / Thurrock area. HMSO1978. ISBN 0 11 883203 4.
71. Canvey. A second report etc. HMSO 1981. ISBN 0 11 883618 8.
72. Treaty of Rome.
73. European Community Act 1972.

74. Single European Act 1986.
75. Redgrave Fife & Machin. Health and Safety Second Edition pp 1xiii ISBN 0 406 02278 X .
76. Report of Chief Inspector of Factories and Workshops for 1880. C-2825.
77. Report of Chief Inspector of Factories and Workshops for 1881. C-3183.
78. Report of Chief Inspector of Factories and Workshops for 1882.
79. Report of Chief Inspector of Factories and Workshops for 1883.
80. Annual Report of the Chief Inspector of Factories and Workshops for 1910. Cd. 5693.
81. Annual Report of the Chief Inspector of Factories and Workshops for 1915.
82. Annual Report of the Chief Inspector of Factories and Workshops for 1916.
83. Annual Report of the Chief Inspector of Factories and Workshops for 1917.
84. Annual Report of the Chief Inspector of Factories and Workshops for 1918.
85. Annual Report of the Chief Inspector of Factories and Workshops for 1919. Cmd. 941.
86. Annual Report of the Chief Inspector of Factories and Workshops for 1920 Cmd.1403.
87. Annual Report of the Chief Inspector of Factories and Workshops for 1921. Cmd.1705.
88. Annual Report of the Chief Inspector of Factories and Workshops for 1922. Cmd.1920.
89. Annual Report of the Chief Inspector of Factories and Workshops for 1923. Cmd. 2165.
90. Annual Report of the Chief Inspector of Factories and Workshops for 1924. Cmd. 2437.
91. Annual Report of the Chief Inspector of Factories and Workshops for 1925. Cmd. 2714.
92. Annual Report of the Chief Inspector of Factories and Workshops for 1928. Cmd. 3360.
93. Annual Report of the Chief Inspector of Factories and Workshops for 1929. Cmd. 3633.
94. Annual Report of the Chief Inspector of Factories and Workshops for 1930. Cmd. 3927.
95. Annual Report of the Chief Inspector of Factories and Workshops for 1931. Cmd. 4098.
96. Annual Report of the Chief Inspector of Factories and Workshops for 1932. Cmd. 4377
    (Including a review of the years 1836 to 1932).
97. Annual Report of the Chief Inspector of Factories and Workshops for 1933. Cmd. 4657.
98. Annual Report of the Chief Inspector of Factories and Workshops for 1934. Cmd. 4931.
99. Annual Report of the Chief Inspector of Factories and Workshops for 1935. Cmd. 5230 .
100. Annual Report of the Chief Inspector of Factories and Workshops for 1936. Cmd. 5514
101. Annual Report of the Chief Inspector of Factories and Workshops for 1937. Cmd. 5802.
102. Annual Report of the Chief Inspector of Factories for the Year 1938. Cmd. 6081.
103. Annual Report of the Chief Inspector of Factories for the Year 1939. Cmd. 6251.
104. Annual Report of the Chief Inspector of Factories for the Year 1940. Cmd. 6316.
105. Annual Report of the Chief Inspector of Factories for the Year 1941. Cmd. 6397.
106. Annual Report of the Chief Inspector of Factories for the Year 1942. Cmd. 6471.
107. Annual Report of the Chief Inspector of Factories for the Year 1943. Cmd. 6563.
108. Annual Report of the Chief Inspector of Factories for the Year 1944. Cmd. 6698.
109. Annual Report of the Chief Inspector of Factories for the Year 1945. Cmd. 6992.
110. Annual Report of the Chief Inspector of Factories for the year 1946. Cmd. 7299.
111. Annual Report of the Chief Inspector of Factories for the Year 1947. Cmd. 7621.
112. Annual Report of the Chief Inspector of Factories for the Year 1948. Cmd. 7839.
113. Annual Report of the Chief Inspector of Factories for the Year 1949. Cmd. 8155.
114. Annual Report of the Chief Inspector of Factories for the Year 1956. Cmnd. 329.
115. The Annual Report of HM Chief Inspector of Factories, 1964. Cmnd. 2724.
116. The Annual Report of HM Chief Inspector of Factories, 1966. Cmnd. 3358.
117. Annual Report, 1972. HM Chief Inspector of Factories. Cmnd. 5398.
118. Annual Report, 1973. HM Chief Inspector of Factories. Cmnd. 5708.
119. HM Chief Inspector of Factories. Annual Report, 1974. Cmnd. 6322.

INDEX

If you are interested in purchasing other books published by Tempus,
or in case you have difficulty finding any Tempus books in your local bookshop,
you can also place orders directly through our website

www.tempus-publishing.com